THE GREAT ACCELERATION

THE GREAT ACCELERATION

How the World is Getting Faster, *Faster*

ROBERT COLVILE

BLOOMSBURY
LONDON • OXFORD • NEW YORK • NEW DELHI • SYDNEY

Bloomsbury Publishing
An imprint of Bloomsbury Publishing Plc

50 Bedford Square
London
WC1B 3DP
UK

1385 Broadway
New York
NY 10018
USA

www.bloomsbury.com

BLOOMSBURY and the Diana logo are trademarks of Bloomsbury Publishing Plc

First published in Great Britain 2016

British Library Cataloguing-in-Publication Data
A catalogue record for this book is available from the British Library.

ISBN: HB: 978-1-4088-4007-8
TPB: 978-1-4088-4008-5
ePub: 978-1-4088-4006-1

2 4 6 8 10 9 7 5 3 1

Typeset by Newgen Knowledge Works (P) Ltd., Chennai, India
Printed and bound in Great Britain by CPI Group (UK) Ltd, Croydon CR0 4YY

To my father, who taught me how to take things slowly.
And to Andrea, who makes my heart beat faster.

CONTENTS

Introduction

'All things move, all things run, all things are rapidly changing.'
– *The Technical Manifesto of Futurist Painting*[1]

Next time you walk down the street, take a look at people's feet. Pretty quickly you will notice that wherever you are, however large the crowd, they are marching in perfect lockstep. You can try to break the pattern – to speed up or slow down – but without a supreme effort, you'll find that you've simply lengthened or shortened your stride: your feet are still marching to that same common beat.

This is, when you think about it, rather amazing. Our natural sense of rhythm is so strong, our inner beat so powerful, that it overrides our conscious thoughts. Even if you're listening to music, it doesn't matter: your iPhone may be pumping out Schubert while your neighbour's plays Jay-Z, but your legs will remain in perfect synch.

The reason we fall into step is due to a phenomenon called entrainment, in which living creatures' natural rhythms unconsciously adjust to each other. It's the same strange force that sees swallows wheel and dive in unison, or brings together the menstrual cycles of women who share the same house. And when it comes to walking, it would be natural to think that this collective pace is ruled by some primal biological rhythm, like the beat of our hearts. But in fact, it is set by our surroundings. In different places, people entrain in different ways: sometimes we are tortoises, sometimes hares.

This pace is determined, above all, by the nature of our environment. The larger the town or city, the faster people move: in the 19th century it was said that the average New Yorker 'always walks as if he has a good dinner before him and a bailiff behind him'.[2] Today, a child brought up in a city will race around a supermarket more than twice as fast as his cousin from a small town, who will spend far more time talking to the staff or inspecting the produce.[3] And if you ask people to listen to a pause of a particular length, those living in cities with a population of more than a million will claim it has lasted twice as long as those from farms or villages.[4]

It is not just the size of communities that matters, however, but their culture. During a teaching sabbatical in Brazil, a Californian psychology professor called Robert Levine found his American obsession with punctuality utterly unsuited to the laid-back local norms. He decided to focus his research on why the pace of life varied around the world: so he and his students spent three years in the early 1990s visiting 31 different cities to measure the differences.

It turned out that the more advanced and industrialised the economy, and the more individualistic the culture, the faster the country's speed. Western Europe and Japan hurried and scurried, while Africa and Latin America dawdled and lazed. Within America, the East Coast was the fastest, followed by the West, while the heartland ambled behind.[5]

Yet cultures change over time – and with them the speed of life. In 2006 a psychologist from the UK called Richard Wiseman enlisted the help of the British Council to carry out a rerun of Levine's experiment. On a given day in August, its staff took to the streets of 32 cities around the world. Like Levine's team a decade earlier, they staked out a 60-foot strip of unobstructed pavement in the heart of the city centre, grabbed a coffee, and started the stopwatches.[6]

What did they find? That worldwide, people were covering the same stretch of ground in 10 per cent less time. In particular, the cities of Asia had caught the speed bug. Levine had been surprised that, despite their reputation for haste, the booming, bustling cities of the East were far less speedy than sclerotic old Europe (due, he theorised, to their warmer temperature). But during the years between the two experiments, Singapore and Guangzhou had come from nowhere to match the hastiest Western capitals – just as the Asian economy had first borrowed the West's hyperactive ethos, and then turbocharged it. In an act of wholesale cultural imperialism, America and Europe appeared to have impressed upon the developing world not only their individualistic, consumerist culture, but also a Western sense of haste.

THE GREAT ACCELERATION

What single quality best defines how our society is changing? Is it that life is becoming fairer, or more equal, or more prosperous? No: as the experiment above suggests, it is that life is getting faster. This is something many of us will have experienced for ourselves. This book, for example, was born out of the realisation that it was not just our electronic devices that were getting faster, nor the pace of our working lives. The media industry in which I worked, and the political system which I covered, were subject to sudden and convulsive change. New trends, ideas and crises appeared to be emerging in the blink of an eye. It felt like my friends and I had no time to relax, to unwind, to slow down.

The more I studied this phenomenon, the more I realised that it was all connected. In area after area, technology was making life quicker, more convenient, more friction-free – not least as more and more of it moved online. The debates people were having about the internet's effects on our brains, or our rapacious use of the planet's resources, were all part of the same basic phenomenon: what I call the great acceleration.

Often, this acceleration feels like something we cannot control. Where our grandfathers sauntered and our fathers strode, we find ourselves scampering about our daily tasks. Time-use studies have shown how work increasingly follows us from the office to the home, with everyone tethered to the office by their smartphone.[7] And that same phenomenon of entrainment – the shared sense of rhythm that sees us all marching in lockstep – means that we instinctively, unconsciously, push ourselves to match the pace of our fastest fellow workers.[8]

In 1990, 49 per cent of Europeans felt their work schedule was too strenuous. By 2000, that had increased to 60 per cent – and those who felt themselves to be rushed were almost twice as likely to complain of the classic stress disorders, such as back pain, or tight shoulders and necks.[9] Bombarded by electronic and bureaucratic stimuli, we are pushed into a near-perpetual state of fight or flight. We are apparently so stressed that the very gender balance of our society has shifted, with more girls being born than boys – a change more normally seen in the wake of traumatic events such as Hiroshima.[10]

The psychologist Stephanie Brown argues that America's frenzied lifestyle 'has itself become an addiction. People are out of control in their push to do more, to always be on and available, and never to say no.'[11] Speed means progress and success. Slowing down means failure and loss. To regain a sense of control, we fall into other addictions – eating, gambling, computers. 'For many people,' she laments, 'their relationship to technology and speed has become more important than, or even replaced, human relationships.'[12]

Yet at the same time, ours is a society in which speed is not just omnipresent, but venerated. Firms in Silicon Valley compete to be the quickest-moving, the most disruptive. A study of Christmas round-robin letters sent since the Sixties shows not only a remarkable rise in the use of words like 'hectic', 'whirlwind', 'consumed' or 'crazy' to describe people's lives, but their being deployed in an almost boastful fashion, as a proclamation of worth.[13] The novelist Jonathan Franzen has written of a society that seems to be becoming 'as restless as capitalism itself'.[14]

As life speeds up, our patience thresholds dwindle. In 1999, websites would lose a third of their traffic if they took eight seconds to load. By 2006, that had shrunk to four.[15] Now, Google puts them on notice if it takes more than two.[16] In a survey carried out by the psychologist Philip Zimbardo among readers of *USA Today*, the level of anger and frustration caused by every kind of annoyance, from waiting in a queue to getting stuck in traffic, had shot up over the past 20 years.[17]

It is not just our biology that is changing: our economy, too, is being altered in ways that many will not find comfortable. The process of disruption and automation – the shift from manual labour to computer power – is throwing industry after industry into turmoil, as first the working and now the middle classes discover that computers can do their jobs faster and cheaper than they can themselves. John Maynard Keynes called this 'technological unemployment' – and the faster technology moves, the less we are able to cope with the dislocation it causes.[18]

Yet the central argument of this book is not only that acceleration is a good thing, but it is something that we have actively chosen. We are not mere passive victims of some vast, impersonal force: we have, collectively, chosen to bring the great acceleration on ourselves. We are, as humans, hard-wired to crave novelty, speed and convenience. Despite all the stress, people from faster places are far happier with their lives. As Levine found, 'faster overall tempos are highly related to a country's economic well-being on every level: to the economic health of the country as a whole . . . to the economic well-being actually experienced by the average citizen . . . and to how well people are able to fulfil their minimum needs.'[19] In other words, those people scurrying through the world's great cities are by and large doing so, as we shall see, with a spring in their step.

Set against a depressing picture of stress and uncertainty, indeed, is an equally powerful positive argument – that in the words of Adrian Wooldridge, management columnist on the *Economist*, 'the storms of creative destruction are blowing us to a better place'.[20] Acceleration has brought prosperity, on a scale unimagined even by the optimists. In 2003, in a paper that coined the term BRIC (Brazil, Russia, India and China) for the rising global players of the future, Jim O'Neill and his colleagues at Goldman Sachs ventured some predictions about how quickly power and wealth would shift from West to East. They imagined that by 2008, China's GDP could be as high as $2.8 trillion; in fact it stood at more than $4.3 trillion. Brazil's GDP, similarly, was two and a half times greater than projected.[21] This ongoing process of economic acceleration has lifted billions of people out of poverty.

Writing in the *New Yorker* in 2011, Adam Gopnik claimed that commentators on technology fall into three basic categories: the Never-Betters, Better-Nevers and Ever-Wasers.[22] 'The Never-Betters,' he wrote, 'believe that we're on the brink of a new utopia, where information will be free and democratic, news will be made from the bottom up, love will reign and cookies will bake themselves.' The Better-Nevers are more likely to be found lamenting the loss of old certainties and expressing their support for UKIP or the Tea Party. The Ever-Wasers, meanwhile, sagely observe that technological acceleration has been a fact of life since the invention of the Bessemer engine, and that human reactions to it have been remarkably consistent throughout – namely, embracing every possible opportunity for greater speed while complaining about it every step of the way.

The debate about whether the acceleration of life is a good or bad thing has, as this suggests, been going on for as long as life itself has been speeding up. Anyone who worries about families studying their iPads or the television rather than talking to each other should consult the edition of the *Journal of Education* from 1907 which complained about 'our modern family gathering, silent around the fire, each individual with his head buried in his favourite magazine'.[23] Similarly, John Freeman's book *The Tyranny of E-mail* outlines a host of past irritations that sound suspiciously familiar to those immersed in today's online culture. Spam? Invented in 1868 by G. S. Smith, who built up a staff of 430 people, all churning out unwanted advertising circulars.[24] Nigerian email fraudsters? Try the British-American Claim Agency of the late 1880s, which informed gullible Yankees that they had received legacies – even landed estates – from relatives in England, which they could claim in exchange for a small handling fee.[25] Abusive online comments? Soon after the birth of the postcard, there were widespread complaints that they were filled (anonymously) with 'gross insolence or contemptuous epithets'.[26]

But just because these arguments are familiar does not mean they are not urgently important. In field after field of human activity, the pace of change is accelerating. If we get dazzled by the potential of technology in the way that some in Silicon Valley have, we will be blind to the very real dangers that a lightning-quick economy and society bring. But similarly, if we focus solely on the negative, we will miss the glorious possibilities that await.

Only the most curmudgeonly of traditionalists could fail to feel a twinge of excitement about some of the technologies that seem likely to be with us sooner rather than later: an 'internet of things' built of 100 trillion sensors; self-driving cars; photo-realistic virtual reality; universal 3D printing; medical nanobots able to diagnose and cure diseases without our even knowing. Conversely, only the blithest of Panglossians could be sanguine about the ever-expanding range of ways in which the human race might imminently doom itself, from our rapacious strip-mining of the planet's resources to the prospect of creating superintelligent machines whose interests we have failed to align with our own.

FUTURE PERFECT?

This is not, primarily, a book about technology itself. Rather, it deals with human nature and its response to the technology around us. Each chapter explores how a different particular aspect of society is being transfigured by acceleration. While others have written powerfully about the speeding up of life – such as James Gleick in *Faster*[27] – this is the first book to draw together the threads, and show quite how pervasive and universal its effects have become.

To that end, the first chapter examines the acceleration of technology, and the way in which ever-greater speed is baked into modern business. The second chapter examines the impact of this on our brains and bodies, and the third its effects on our social and romantic lives – how we live, work and raise our children. Chapter 4 traces how acceleration is shaping popular culture, while chapters 5 and 6 examine its disconcerting effects on the media and politics. Chapters 7 and 8 turn the spotlight on to the financial markets and the global system of trade and logistics, while Chapter 9 examines the effect of all this rapacious activity on the natural world. The conclusion explores where this process may ultimately take us – to a world of artificial intelligence, artificial life and even artificial humans.

What emerge from this are seven common themes, which crop up again and again.

First, the benefits of the great acceleration are more substantial than its costs, but less dramatic. We are geared to pay more attention to the one day on which the stock market drops by 10 per cent than the long months in which it ticks slowly up. We worry about the impact of mobile phones on children's brains, without appreciating the miraculous convenience of the connectivity they provide.

Second, the quickening pace of life is promoting the flashy and the superficial. But there is also an expanding market for products and services that help us step out of the currents and catch our breath, such as box sets, long reads or spa treatments. What is suffering instead is what sits between fast and slow: products or experiences that are neither instant nor immersive, but in the mushy middle. For consumers, this has been a very good thing – even if it has made life distinctly uncomfortable for many producers.

Third, this polarisation between fast and slow is being mirrored by a polarisation between large and small. Whether it is in the technology industry or publishing or popular culture, the great acceleration has created a horde of buzzing, disruptive innovators. But it has also created and rewarded a few giants who have mastered the business of speed, flourishing and dominating the same ecosystem as those of small, nimble operators. This is a world not of hyenas, but of lions and flies, with those who are neither giant generalists nor niche specialists increasingly squeezed out.

Fourth, the benefits of faster and faster change come with a price tag of less and less predictability. An accelerated world is an exciting place, but is also one characterised, as Karl Marx put it in the *Communist Manifesto*, by 'the constant revolutionising of production, uninterrupted disturbance of all social conditions [and] everlasting uncertainty and agitation'.[28]

Fifth, that price tag also includes fragility. When the pace of change accelerates, we have less leeway to adapt, meaning there is more chance of things going catastrophically wrong. Our just-in-time logistics systems give us cheaper goods and cheaper food, but leave us vulnerable to shocks and disruption. Moving services online makes them easily accessible, but also vulnerable to hacking. Similarly, the ease with which money, ideas and pathogens can spread around the world – the lack of friction in the system – means that disaster can spread before we are even aware of it, let alone in a position to respond.

Sixth, the trends that power the great acceleration all feed off each other. We are drawn to live in cities for their speed; living in cities makes us more innovative and productive; this generates better technology, or new ideas; these make our lives faster still and increase our prosperity; as we become more prosperous, we become more urban – and so on. Such feedback loops arise time after time, each one speeding up the great acceleration.

Finally, this new world is not necessarily fair. In the words of the racing driver Michael Schumacher, 'to perfect things, speed is a unifying force. To imperfect things, speed is a destructive force.'[29] Those people or institutions that are worst placed to respond to acceleration will suffer more from its side effects, while those willing and (crucially) able to adapt to it will make outsize gains, for example from faster devices, or biological enhancements that allow them to think better and act faster.

In the aggregate, the great acceleration is an extraordinarily good thing for humanity. But its benefits are distributed unevenly, and its dangers are almost as great as its opportunities. That is why understanding how to blunt the worst consequences of change and embrace its best effects has never been more important. We cannot stop this acceleration: what we can do instead is ensure that its enormous potential is applied where it best serves our needs. We have it in our hands to build the greatest and most prosperous society in history, or to wreck ourselves through selfishness and greed. Which path we take will be determined by whether we become the slaves of the great acceleration, or its masters.

I

Permanent Revolution

'Anything worth doing is worth doing faster.'

– Advertising slogan for the BlackBerry Playbook

'Google is dead.' Bill Nguyen makes this pronouncement casually, almost offhandedly – not as if he's handing down a truth, but mentioning something that's already common knowledge.[1] What, you guys didn't know already?

The date is May 2011, and the dapper, friendly Vietnamese–American is the toast of the technology industry. Having founded a string of successful firms before the age of 40, the most recent snapped up by Apple, he has just secured $41 million in funding for his new creation, Color, a location-based photo-sharing website. It is the subject of a tidal wave of hype – being touted as 'miraculous' and 'transformative', the new new thing that could knock Google and Facebook off their perch.[2] The fact that he has agreed to drop by the *Daily Telegraph*'s London headquarters to address its reporters is seen by the management as a significant coup.

At the most basic level, what Nguyen means is that the web is shifting away from the automated model pioneered by Google and towards the social approach favoured by Facebook, in which your experience of the internet is filtered by that of your friends. But he is also talking about something more nebulous: the idea that the stardust has vanished, that Google is no longer the coolest kid on the block. Younger, hipper, faster firms are gearing up to take its place. Firms like Color.

Fast-forward just a few weeks, and it is clear that Color is a flop – and a spectacular one. In a startling about-turn, Nguyen is forced to admit that the network of users he was hoping to create is already in place over at Facebook, and has no desire to move. In the months after its humiliating failure to launch, Color is rebuilt from the ground up, to work as just another app within Facebook's sprawling ecosystem. Within 18 months, the firm will have shut down completely.

Is this just a tale of executive hubris – a cautionary fable about just how deluded the novelty-crazed investors and entrepreneurs of Silicon Valley can be? On the contrary. It tells us some extremely important things about the culture of innovation and disruption that dominates the business world. It is a culture, as we shall see in this chapter, which fetishises ambition, rapidity and transformational, world-changing ideas – in which the rewards for success are greater and the costs of failure lower than ever. And it is one which is built, above all, around acceleration: the acceleration of technology, of business, and above all of customers' tastes and demands.

CHIPS WITH EVERYTHING

Color was, like many technology companies, built around speed. The idea behind its app was that, using just your mobile phone, you would be able to take a high-quality digital picture and have it pop up, within an instant, on the screen of a friend sitting a continent away.

That we now find this idea relatively mundane does not make it any less extraordinary. For it relies on computing power and data transmission on a scale beyond anything available to previous generations.

This is the most obvious factor powering the great acceleration: the way in which the power of computing devices has grown so swiftly. The most famous example is Moore's Law. In 1965, the co-founder of the chip-maker Intel, Gordon Moore, observed that the number of transistors that could fit on a single chip had been doubling every two years (or every year, in the original formulation), and would continue to do so. For the computing industry, this became something of a self-fulfilling prophecy: over the past two decades, global computing power has grown by 58 per cent a year.[3] This is why a PC today costs a sixth of the amount it did in 1981, but is 500 times more powerful – and why your fridge probably has more computing power than the Apollo 11 module.[4]

It is not just raw computing power. The American futurologist and technologist Ray Kurzweil has shown that computing power, data transmission, memory storage – pretty much any technological metric – is subject to what he calls the 'law of accelerating returns', under which its growth is not linear, but exponential.[5] Imagine, he says, that the emperor of China wishes to reward the man who came up with the game of chess. The wily inventor asks for an exponential reward: one grain of rice on the first square of his chessboard, two on the second, four on the third, eight on the fourth and so on. As the sacks are brought in, the emperor thinks he is getting the better of the bargain. But slowly, the exponential curve takes effect. By the final square, the emperor's generosity will cost him 2^{63} grains of rice, or 9,223,372,036,854,775,808. At a rough approximation, that works out to 250 billion tons, or 40,000 times the weight of the Great Pyramid at Giza.

The power of Moore's Law, and of the law of accelerating returns, can be hard to appreciate. 'It's as if you kneel to plant the seed of a tree,' says the technologist Jaron Lanier, 'and it grows so fast that it swallows your whole town before you stand up.'[6]

One of the best places to see this phenomenon in action is CERN in Geneva – famous among scientists as the home of the Large Hadron Collider, and among techies as the place where Tim Berners-Lee created the World Wide Web (if you look hard enough, you will find a plaque commemorating the achievement outside his old office, opposite an outdated collage of Dilbert cartoons).

The reason Berners-Lee was at CERN in the first place was to facilitate information-sharing – not least between scientists in different parts of the world. In those days, that was achieved via a link part-funded by IBM, running at a then-astonishing 1.5 megabits per second. Fast-forward 20 years, and the humblest 3G smartphone can receive data almost ten times faster.[7]

But CERN hasn't stood still, either. Within the Large Hadron Collider, an $8 billion particle collider designed to mimic conditions a millionth of a second after the Big Bang, beams of particles race along a 17-mile circular track at 99.9999991 per cent of the speed of light. When they collide, at a billion kilometres per hour, they throw off vast amounts of data – too much for any one supercomputer. Instead, the work is divided between thousands of machines in hundreds of data centres worldwide, in a network known as the Grid.

With its ability to transfer billions of gigabytes per year, the Grid is the bullet train to the internet's locomotive. But after another 20 years of Moore's Law the processing power available to all of us will be a million times greater. Sooner than we realise, we will all have access to practically infinite data, and practically infinite processing power, at practically infinite speeds.

The availability of all this ever more powerful hardware is one of the most important forces behind the acceleration of society. And the possibilities it offers are dizzying. Within five or six years, Intel plans to shrink its fastest processors from 14 nanometres down to 5, a scale so tiny that quantum physics – notably, our inability to pin down both the exact state and the position of subatomic particles – starts to become an actual design factor.[8] Beyond that lies the point where you can construct processors – or pretty much anything else – molecule by molecule. That has the potential to revolutionise not just computing, but everything from construction to medicine to energy to agriculture, via the construction of nanotech devices and materials.

A QUICK HISTORY OF DISRUPTION

This acceleration of hardware is necessary for the great acceleration to take place – but it is not sufficient. There have, after all, been convulsive episodes of technology-driven change in the past: mass electrification, say, or the reshaping of cities around the car in the post-war years. What makes things different now?

The answer to that is not a question of devices, but doctrines – and, in particular, our approach to innovation.

There have always been two basic recipes for innovation: making an incremental improvement to a product, service or process, or coming up with something brand-new. This is the distinction made by Larry Page, co-founder of Google, when he says that his firm's aim should be not to improve existing products by 10 per cent, but to build ones that are ten times better.[9]

A classic example of this kind of disruptive innovation comes from the communications revolution of the 19th century. As late as 1845, it took President James Polk six months to get a message to California.[10] Yet by the time of Lincoln's inaugural address in March 1861, that had been cut to seven days and 17 hours.[11] This was partly thanks to better roads and railways – themselves a huge engine of social and economic change – but also to the efforts of the newly formed Pony Express, which based its business model on detaching its horses from cumbersome stagecoaches and flogging the poor beasts to exhaustion (and its riders, too, as its recruitment ads made clear: 'Wanted: Young, skinny, wiry fellows not over

18. Must be expert riders, willing to risk death daily. Orphans preferred').[12]

The Pony Express was a marvel of logistics – not to mention bravery, given the hostile terrain its young riders would pass through. But within just two years it had folded, victim of the most fundamental law of the market: that the fast displaces the slow.

What eclipsed the Pony Express was what the *Sacramento Bee*, in its obituary for the service, described as a 'new and higher power': the telegraph.[13] This was an explosive, disruptive and enormously influential technology – what Tom Standage of the *Economist* calls 'the Victorian internet'.[14] At the start of 1846 there was only one experimental 40-mile line, laid between Washington and Baltimore by Samuel Morse.[15] By 1850, 12,000 miles had been laid.[16] By 1858, the first transatlantic cable had entered service, to prodigious rejoicing. ''Tis done!' reported one anonymous poet. 'The angry sea consents, / the nations stand no more apart; / With clasped hands the continents / Feel throbbings of each other's hearts.'[17] So vigorous were the celebrations that City Hall in New York was set ablaze, and only just escaped being burnt down.[18]

It wasn't just the telegraph. Just as we do today, the Victorian public seized on anything that offered greater speed or convenience. The first postcard was sent in 1871; by 1873, more than 72 million had been dispatched.[19] In 1896, the first motion pictures were projected for the viewing public in a New York music hall; by 1910, despite the best efforts of Thomas Edison's patent lawyers, the nascent movie industry was turning out 200 one-reel films a week.[20]

The same process can be seen in transport. Today, footage of Edwardians wobbling around on penny-farthings is a shorthand for sepia-toned gentility. Yet the bicycle was, in its day, a piece of cutting-edge technology – responsible, at one stage, for a third of patents granted in the US, as manufacturers raced to improve the design.[21] The greatest sign of its influence is that it spurred the same moral panic as every transformative technology from Facebook to steam locomotives: there were dark warnings of the sinful consequences of unmarried couples taking rides in the country together, and of the dangers of 'bicycle face', suffered by women who tried to pedal against the breeze at high speeds.[22]

It was developments like this that led the great Austrian economist Joseph Schumpeter to come up with his theory of 'creative destruction'.[23] He argued that waves of disruptive change were not merely a part of business and technology, but their very purpose, sending out ripples of innovation that made markets work better and consumers more satisfied.

So why can we talk about a great acceleration, rather than a 'great continuation'? Because this kind of change is speeding up. Ray Kurzweil has calculated how long various technologies have taken to win mass acceptance (which he defines as use by a quarter of households). For the printing press, it was centuries; for television and radio, a few decades; for the web, just six or seven years.[24] The curve, as Kurzweil points out, is getting ever steeper, meaning that each new technology is embraced more and more quickly. In 2005, just 5 per cent of American adults used social networking sites; six years later, the figure was at 65 per cent and climbing fast.[25]

What's changed? In the old days, the public embraced new technologies such as postcards and moving pictures – or the radio, or the railroad – just as eagerly as we do today. But such innovations faced a rockier journey to the mass market, due to obstruction by incumbent firms or a failure of vision among unimaginative investors.

As early as 1816 – 30 years before Samuel Morse – an Englishman called Francis Ronalds demonstrated a working telegraph over 8 miles of wire strung up in his back garden, but the Admiralty expressed no interest, and there was no early venture capitalist to step in instead.[26]

Lack of imagination was often compounded by sabotage from incumbents: as Machiavelli argued, 'the innovator has for enemies all those who have done well under the old conditions, and lukewarm defenders in those who may do well under the new'.[27] Lawyers at AT&T, the industrial conglomerate created by Alexander Graham Bell, managed to delay the introduction of the voice recorder, the television and the fax machine by years in order to protect their existing business.[28] Even breakthroughs such as radio and film were soon colonised by corporations. Tim Wu, the Harvard academic and author of *The Master Switch*, calls this 'The Cycle' – a process in which innovation and openness first arise, then are remorselessly choked off by cartels and monopolies.[29]

Today, it is certainly true that incumbents do their best to maintain their position: witness the way in which the big car companies sought to blunt the impact of electric vehicles until Elon Musk's firm Tesla came along to shock them into action. But for all manner of reasons, this obstructive mentality is much harder to pursue successfully than it was.

Executives today, surveying the turbulent economic and technological landscape, are all too aware of the dangers of complacency, or clinging to a broken business model too doggedly. The result is a commitment to self-disruption which has seen the concept not just embedded in 21st-century capitalism, but fetishised to the point of near-absurdity – resulting in a culture of which Bill Nguyen is the personification.

This mentality was already growing in power, but the man who gave it the status of holy writ was Clayton Christensen, the American academic and business philosopher whose 1997 book *The Innovator's Dilemma* has become the bible of modern capitalism.[30] In it, Christensen tells the story of the steel industry, traditionally centred on vast industrial foundries. Starting in the 1970s, smaller 'mini-mills' appeared that could process steel more cheaply and efficiently. The market leaders weren't worried: the mini-mills could only make the worst quality of steel, and abandoning that highly competitive market would raise their profit margins and allow them to concentrate on higher-tech products. Then the mini-mills got good enough to make the next-highest grade of steel. Again, the big firms abandoned this tough, unrewarding market segment, and watched their margins and share prices soar. Eventually, they realised they were being squeezed out of the market altogether – but by then it was too late.

In the AT&T days, the moral of this story would have been to stamp out your enemies wherever they appear, no matter how small. But Christensen's message was to disrupt yourself, lest you be disrupted. And he could point to an elephants' graveyard of former market leaders to drive home his point: for the steel mills, read Kodak, or Borders, or Ashton-Tate (once the third-largest software firm behind Microsoft and Lotus). Christensen's chilling point to his readers in the boardroom was that, in a disruptive age, even good executives, doing sensible things, can still find their businesses sinking beneath them if they are not constantly willing to reinvent what they do. And half-hearted efforts are not good enough, since the threats are coming from so many different directions.

The evidence backs him up: between 1956 and 1981, an average of 24 companies dropped out of the Fortune 500 list every year. Between 1982 and 2006, the figure had risen to 40.[31] A similar study found that a firm was falling out of the rival S&P 500 index every two weeks.[32] In just the last few years, companies such as Nokia, BlackBerry and Zynga have gone from world-beaters to has-beens after failing to predict where acceleration would take their industries.

Amid such a frenzy of creative destruction, the most successful companies have to engage in a constant process of reinvention and internal competition to remain ahead of the pack: no less a firm than Intel credits Christensen for helping to save it from corporate obsolescence.[33] When a consultancy recently asked hundreds of executives to rate their firms for a Global Speed Survey, those companies that self-identified as 'faster' had 40 per cent higher sales growth and 52 per cent higher operating profit.[34]

The result is a mindset in which permanent revolution is not just a business model, but almost the only one that makes sense. To stop their firms growing fat, old and lazy, or being rendered an antique by some bright spark in a Californian bedroom, incumbents are adopting the ethos of a start-up, trying to disrupt themselves before they are disrupted in turn. Gary Hamel wrote in the *Harvard Business Review* back in 1999: 'Face it: out there in some garage an entrepreneur is forging a bullet with your company's name on it. You've got one option: you have to shoot first.'[35] Or, to use Google's unofficial internal motto: 'If you're not fast, you're fucked.'[36]

SCHUMPETER'S CHILDREN

This transformation in business mentality hasn't happened because of enlightened self-interest. It's happened because companies often don't have a choice. And the most obvious reason for this is the opportunities opened up by the acceleration of technology.

Consider, for example, the story of the most important video cassette in consumer history. The tape in question was a copy of Ron Howard's film *Apollo 13*, and it was rented in 1997 by Wilmot Hastings Jr, a software entrepreneur and former Peace Corps volunteer. At the time, Hastings, known to all by his middle name of Reed, was having a few problems, both with his career and his family life. For various reasons, he failed to get round to returning the tape for six weeks, by which time the late fee had grown to $40.

For most people, that would have been the end of it: perhaps, at most, they would have resolved either to return their videos on time in future, or to boycott the vendor. Instead, the aggrieved Reed Hastings decided to build a new company, one that offered unlimited DVDs, delivered by post, with no late fees. (As it happens,

Hastings's co-founder and others dispute this version of events, but it has still attained the status of corporate creation myth).[37]

The business model of this firm – which was given the futuristic name of 'Netflix' – wasn't entirely original, and it certainly wasn't as speedy as walking into Blockbuster. But what marked it out was both its convenience (you didn't have to walk anywhere, simply queue up your film orders online) and its determination to innovate. In 2006 it offered $1 million to anyone who could improve its engine for recommending films to users by 10 per cent – a challenge accepted by teams of computer scientists from around the world.[38] Next, and more momentously, it decided to switch its business model from the physical to the virtual, by streaming or downloading films directly into its users' bedrooms or living rooms.

The results were astonishing. Digital video rapidly came to dominate the home video market, turning the old incumbent, Blockbuster, into just another BlackBerry or Kodak. Today, Netflix is a corporate gorilla, beamed into millions of homes worldwide and raising billions to fund its future expansion, which includes making its own movies and TV shows starring some of the biggest names in Hollywood.

This success was emblematic of the modern business environment in more ways than one. First, Hastings benefited from the accelerating pace of technological progress: the fact that it suddenly became technically feasible to beam movies to our laptops rather than etching them on to magnetic tape or plastic discs. Second, he profited from the general shift from hardware to software – the fact that innovation today is more often about writing code than plugging in wires.

Software innovation has a quicker impact (and is loved by investors) because it is much easier to scale up a product that does not involve building factories, or moving physical objects from place to place. Especially since the marginal cost of adding a million more customers, once your code base is in place, will be no more than the cost of renting the extra server capacity.

The transition from hardware to software also makes it vastly easier to start and grow a company – which, in turn, has accelerated the process of disruption. As John Palfrey and Urs Gasser argue in their book *Born Digital*, 'it's much more cost-effective to become a creator of digital content than it was to create similar works in previous areas. In the 1980s, for example, a start-up rock band would have spent roughly $50,000 to purchase or rent the necessary recording equipment to make an album. Today, a band simply needs a laptop computer and some additional pieces of hardware and software, which might cost less than $1,000.'[39]

In the absence of the need to build physical products, the only costs for today's entrepreneurs are, as the technology journalist David Kirkpatrick put it in his history of Facebook, 'servers and salaries'.[40] And even those costs are coming down, not least since others – especially the Silicon Valley giants – are more than happy to do the donkey work for you. One of the drivers of the recent start-up boom, for example, has been Amazon Web Services, a cloud-computing platform built by Amazon that enables anyone to rent as much or as little computing power and data storage as their business requires. It is now possible, thanks to the acceleration of computer power, to outsource not just your IT, but your financial reporting, your HR systems and much else. If your start-up needs legal services, for example, you can use software called EDiscovery. In 2011, it analysed 1.5 million legal documents for BlackStone Discovery of Palo Alto for less than $100,000 – a task which, done by humans, would have been an order of magnitude more expensive, and significantly less accurate.[41]

The entire Silicon Valley ecosystem, in fact, is now dedicated to nurturing and supporting small teams of people with big ideas. The latest craze is for the 'lean' start-up – start small, create a minimum viable product, and be prepared to switch tack at a moment's notice (a business philosophy best summed up, Bill Nguyen-style, as 'carry on failing until you succeed'). This kind of operation is obviously a lot easier to set up and fund than something that takes a team of hundreds to pursue a particular vision. For the latest start-ups, actual employees or office spaces are something of a hindrance. For companies such as Uber, smartphones turn thousands of people into contractors without adding to their own overheads, with apps enabling such remote workers to receive tasks from a central hub.

The paradox here is that while these start-ups are smaller than ever, their ambitions are far grander. Indeed, this more than anything is why the gospel of disruption has had such an outsize effect on our lives – why it has fostered that 'constant revolutionising of production' that Marx predicted.[42] In this new world, it is no longer enough to carve out a profitable niche (not least because others may come along and disrupt you out of it). To be among the cool kids, you need to be planning to completely upend a particular industry or business practice. When Marc Andreesen, the co-author of the world's first web browser turned venture capitalist, claims that 'software is eating the world', he is talking about precisely this: the way that established market sectors are falling like dominos before the power of Silicon Valley, due to the lowering of barriers to entry enabled by the shift from hardware to code.[43]

Such venture capitalists are, in fact, another engine of acceleration. As Tad Friend put it in a *New Yorker* profile of Andreesen, 'venture speeds the cycle of American impatience: what exists is bad and what replaces it is good – until the new thing itself must be supplanted'.[44] The more ambitious your chosen target, the greater the scale of the market opportunity: hence Uber's plans to revolutionise not just taxi rides but the very concept of car ownership, or the various start-ups and laboratories working to reinvent agriculture or conquer mortality itself. As Andreesen told Friend, 'we're not funding Mother Teresa. We're funding imperial, will-to-power people who want to crush their competition. Companies can only have a big impact on the world if they get big.'[45]

It is no wonder that, as a result of all this, the start-up scene is undergoing what the *Economist* has called 'a Cambrian moment' (a reference to the explosion of multicellular life around 540 million years ago). 'Digital start-ups,' it said, 'are bubbling up in an astonishing variety of services and products, penetrating every nook and cranny of the economy.'[46] The acceleration is not limited to software, either: new technologies such as 3D printing enable companies making actual goods to prototype their products, refine them and bring them to market far more rapidly.

In summary, a series of factors have converged to make innovation and acceleration of business models both the norm and the expectation – whether it be start-ups attempting to eat the incumbents' lunch, or incumbents desperately attempting to fend them off. And, given how thin many companies' profit margins are, in an ultra-competitive global market it takes only a little bit of disruption – the growth of Airbnb as a rival to hotels, for example – to throw things into chaos. Or, as Schumpeter would put it, to inject an element of creative destruction.

THE CUSTOMER IS ALWAYS RIGHT

This business model, while extremely powerful, is not without its problems. For one thing, it breeds an ethos that holds that it is better to act first and ask permission (or forgiveness) later. Better, for example, to release a project mashed together in a week of caffeine-fuelled coding than a more polished product that fails to catch the Zeitgeist. Hence Facebook's all-night 'hackathons', or the motto Mark Zuckerberg donated to his company: 'Move fast, and break things.'[47]

In the software market, the race to stay ahead of the game therefore means springing redesign after redesign on your users.

Sometimes you'll get it wrong – but that's tough luck. As Reid Hoffman, the billionaire founder of the professional social network LinkedIn and an early investor in Facebook, has said: 'If you're not embarrassed by your first version, you waited too long to ship it.'[48] (Hoffman is also the author of the perfect image to sum up the new-technology business model: 'You jump off a cliff, and you assemble an airplane on the way down.')[49]

From the outside, however, it can feel like change is being shoved down users' throats. Douglas Coupland, author of *Microserfs*, recently observed: 'It feels wistful to imagine a time when people didn't go about their daily routine with the assumption that at any moment another massive media technology will be dumped on us by some geek in California.'[50] The same applies to those being competed out of business: one San Francisco driver complained during the middle of the price wars between rival ride-hailing apps Uber and Lyft: 'I don't want to be the butt-end of a [venture capitalist] fantasy, I want to make a living.'[51]

One rejoinder to such complaints is to quote Schumpeter's observation that 'economic progress, in capitalist society, means turmoil'.[52] More turmoil, by this analysis, means more progress. But the best explanation for why we are being constantly bombarded with more innovation is simple – because we asked for it.

Uber and Airbnb, for example, have succeeded not simply because of venture capitalist support. They are successful because enough of us like what they offer. It may take plenty of failed experiments to figure out exactly what we're after – but the combination of faster computers and better data means that this kind of efficient response to user feedback allows such firms to improve their offerings much more quickly than their predecessors ever imagined.

Of course, there has been a longstanding effort on the part of business to squeeze more efficiency out of its operations. This process was kick-started in the 1890s when the first time-and-motion expert, Frederick Winslow Taylor, set to work at the Midvale Steel plant in Philadelphia, carrying out laborious tests to calculate how long it took workers to do everything from open a drawer to pick up a pencil.[53] The equation was simple: greater speed and efficiency equals fewer staff and higher productivity.

Today, the science of management, productivity and leadership is a business in itself: managers carry out constant assessments of their subordinates' performance, armed with all manner of metrics. But it has also become far easier to hone not just how your products are made, but how they are received.

Take the practice of A/B testing. This involves showing two different versions of a product to different people (or, in one famous experiment by Google, 41 different shades of blue on the taskbar).[54] You can get real-time feedback on their preferences, stripping away the need for human judgement, focus-grouping or lengthy product development. Given a large enough sample size, you can simply spit out hundreds or thousands of different versions of your product until you have reached the ideal version.

This is a hugely powerful technique. The Obama presidential campaign used it to tinker with the photos and slogans on its homepage, increasing the number of supporters signing up by 40 per cent (leading to an extra 4 million email addresses and $75 million in donations).[55] When I was an editor at BuzzFeed, we used it on practically every story, creating as many as a dozen different combinations of picture and headline to see which the audience preferred. When you play a game on your mobile phone, you are subject to constant A/B testing: indeed, software companies such as Wooga or Zynga are essentially data analysts masquerading as games designers. The rough launch versions of games such as *Brain Buddies* or *Monster World* are remorselessly honed and refined, made more and more addictive, until people are clicking faster, staying longer and spending more. And if it turns out that this kind of incremental innovation is not doing the job, that there really is no market for what you are selling, you can simply ditch that particular product or strategy – 'pivot', in the jargon – and try something else.

But there's another thing about this kind of testing: namely, that speed always wins. Consider the history of Google.[56] For all the innovation of its web-crawling algorithm, and all the fields it has subsequently diversified into, its original offering to customers was blindingly simple: speed. For the sluggish search engines that dominated the scene before its arrival, the point was to keep you hanging around on their site; for Google, the point was to get you to the site you actually wanted. The plain white homepage was a statement of technological as well as aesthetic intent: fewer images to load meant faster delivery of pages, which meant less strain on the servers, quicker searches and happier customers.

Today, Google is near-evangelical about the virtues of speed. A campaign called 'Every Millisecond Counts' urged its clients to follow its lead and streamline their homepages – with those burdening the customer with slow load times getting penalised in the company's search rankings.[57] One of the firm's public 'guiding principles' is that 'fast is better than slow': 'We may be the only people in the world,' they boast, 'who can say our goal is to have people leave our homepage as quickly as possible.'[58]

In June 2009, Google revealed the results of a fascinating controlled experiment: slowing down its results had 'a measurable impact on the number of searches per user'.[59] The longer the delays, the fewer the searches. According to Vic Gundotra, vice-president of engineering, 'Our internal data shows that users clearly, even subconsciously, prefer sites and applications that are snappier.'[60] In 2007, Ron Kohavi and Roger Longbotham of Microsoft revealed that a tenth-of-a-second increase in the loading time of Amazon.com cut sales by 1 per cent, while a half-second delay in displaying Google results slashed revenue from that page by 20 per cent.[61] And more recently, a senior executive at the search giant confided that improving loading times by as little as 400 milliseconds raised traffic by 0.5 per cent.[62]

In other words, it's not just that we demand speed: we demand ever greater speed. This is why Google now finds search results for you while you are still entering your query, or Apple pops up suggested words as you enter your text messages. It is only by being 'faster than the speed of type', they believe, that they can satisfy their ever more impatient users.[63]

The history of another Silicon Valley giant shows the same pattern. Its story may be familiar: a brilliant maverick programmer, broken-hearted after a bad break-up, decides to create a social networking site, essentially to meet girls. There have been a few such sites before, but his is different. For a start, you have to use your real name, forcing a certain level of accountability and maturity upon the users. There is also a 'killer app' – the ability to easily upload photographs of yourself from your newfangled digital camera.

The site goes live in March 2003; within six months, 3 million users have flocked to it. Its founder has, in effect, discovered a way to speed up the process of friendship, and has become a superstar in the process: there are interviews, magazine covers, investment from the venture capital elite. Within a year, Google weighs in with a $30 million buyout offer, but it is turned down.

But here's the twist. What I am describing isn't the creation of Facebook, but of Friendster, an earlier social network founded by a Silicon Valley hotshot called Jonathan Abrams.[64] It had everything that Zuckerberg's creation later would: the business model, the backing, the audience. So why haven't you heard of it?

The answer is simple: speed. Amid the excited discussions about new markets and features, and squabbles between founder and board, the team at Friendster forgot to focus on making the site work. Users suffered crippling loading delays as demand overwhelmed capacity. Soon they were deserting in droves. Although the site retains a foothold in parts of Asia, Friendster is now, as the *New York Times* has said, Silicon Valley shorthand for unmet potential.[65]

So when, in February 2004, Mark Zuckerberg launched The Facebook (as it then was), his most important asset was not his skill at coding, or his Harvard connections, but – as David Kirkpatrick says in his history of the company, *The Facebook Effect* – a burning desire not to be 'Friendstered'.[66] Zuckerberg's site, just like Larry Page and Sergey Brin's, stripped away everything extraneous. He scorned rivals that were, in his words, 'too useful', that just did 'too much stuff'.[67] One of his biggest arguments with his initial business partner Eduardo Saverin came when Saverin wanted to add an extra click to the process of requesting a new friend, in order to shove another advert in front of the user. For Zuckerberg, this was 'apostasy'.[68]

Everything Facebook has learned since bears out the wisdom of this approach. When it first brought in the News Feed, allowing you to monitor your friends' updates in real time, the main protest group attracted 700,000 members within three days. Zuckerberg apologised, and tweaked the privacy settings. But the News Feed stayed, for one simple reason. Before the change, users were looking at 12 billion pages a month. Soon after, the figure was 22 billion.[69]

In 2008, when Zuckerberg combined the Wall (the user's homepage) with the News Feed, the explicit rationale was 'to increase the velocity of information flowing between users'.[70] And when Facebook expanded its Messages software to bring together its users' email accounts, text messages and chats, it was, said Zuckerberg, because its teenage users were abandoning formal emails – with their cumbersome need for subject lines and proper spelling – for even-more-instant messages, just as they have abandoned voice calls in favour of snappier texts.[71]

Far from being shoved down our throats, therefore, acceleration turns out to be something we crave – and demand. We will explore the neurological basis for this in the next chapter, but for now suffice to say that its effects are both powerful and inexorable.

One final example will prove the point. For years it was claimed that piracy would destroy the music industry, and the software business too. It is certainly commonplace: in 2011, the British music industry estimated that more than three-quarters of the songs downloaded in the UK were illegally obtained, with twice the proportion of 16- to 54-year-olds stealing music as regularly buying it (29 per cent as against 14 per cent).[72] In 2009/10, as the Kindle and its cousins entered the mainstream, there was a 50 per cent rise in Google searches for unlicensed copies of books.[73] The Business Software Alliance once calculated that P2P traffic – the transfer of files, usually illegal ones, directly between computers – consumed anywhere between 49 per cent and 89 per cent of all internet traffic, rising to up to 95 per cent at night.[74] Bill Gates has claimed that fewer than a tenth of a per cent of the copies of Windows on sale in China are genuine.[75]

But if piracy is a problem, what kind of problem is it? Is it a problem of culture, or morality, or technology? Or is it a problem of speed? It turns out, it is very definitely the last of these. In June 2015, an exhaustive UK government report found that three in five internet users have downloaded or streamed films, games, books, music, TV shows or software – but only one in five had done so illegally, and of those a significant proportion said they only did it because the content they wanted was not available quickly or legally.[76] In the US, the proportion of people using illegal file-sharing services to download music has been cut in half.[77]

So why has piracy fallen – or, at the very least, why has it failed to destroy the music, entertainment and software industries at quite the rate that was predicted? In a word, speed. People downloaded music illegally because it was the simplest and speediest way to do it. Once iTunes came along, it became far easier to play by the rules – and even more so once iTunes was itself disrupted by the even more convenient streaming services available on YouTube or Spotify.

Or look at the success of Apple's App store. Technically, it has long been possible to bypass its security. But the speed and convenience of tapping on a symbol and watching your app download have trumped the relatively minor investment of time and effort it would take to crack the system open.

The logic of the on-demand, instant-gratification economy is that consumers will reward services that offer them what they want, when they want it, and punish those that don't or can't. This doesn't just apply to start-ups: what are customer reviews on TripAdvisor but a powerful spur for hotels or restaurants to clean up their act and offer the best possible service? In the case of video streaming, as the US technology journalist Farhad Manjoo noted:

> Netflix's dominance over BitTorrent [one of the main file-sharing services, whose traffic was rapidly eclipsed by Hastings's creation] fits into a larger story about how our Internet use is changing . . . We're using more of our bandwidth to download stuff we need right now, and less for stuff we need later . . . once we come to expect immediate access to videos, BitTorrent's download-now, watch-later model seems outdated.[78]

THE REVENGE OF THE DINOSAURS

The corporate landscape described above – with companies pushing to disrupt and innovate, and customers pushing them on further – would seem to be a recipe for perfect competition. Nowhere should that be more apparent than in the technology sector, which is where many of the effects of the great acceleration not only originate but also are most apparent. If the fast is indeed eating the slow, shouldn't the giant firms be dying off and making way for smaller, nimbler creatures?

But this is perhaps the most counter-intuitive consequence of the accelerated business culture – and a phenomenon we will run into again and again. The advantages of moving fast are not evenly distributed. Instead, they tend to build on themselves, to the point where those who get ahead of their rivals will end up with a dominant, even monopolistic position.

There are several reasons for this: such firms have the systems in place to scale up faster; their larger size allows them to drive down prices for their customers; and network effects mean that, if their products have a social component, they become more useful the more people are using them (think Facebook or Twitter or even TripAdvisor). This is not just a Silicon Valley phenomenon, either: the momentum towards gigantism is everywhere in the modern market, propelled (as we shall see later) by the financial markets. Investors demand that companies hit endless short-term growth and revenue targets, which they can often do only by scaling up via acquisition.

So why has this not slowed the pace of innovation, in the way that Tim Wu's 'Cycle' predicts? Well, in places it has: as mentioned earlier, the automotive giants showed little interest in electric cars until they were disrupted by Tesla. But there is something qualitatively different. For all the reasons mentioned above, the new breed of monopolists is devoted to sustaining innovation, not least since many of them emerged from acceleration and are committed to pushing it on. Indeed, they frequently compete to host their own innovative parasites: companies or individuals making games or goods or videos that they can distribute via their own channels.

Perhaps the best example of all these trends is Amazon, founded by Jeff Bezos (a former employee, not coincidentally, of one of the computer-driven stock-trading firms whose work we shall examine in chapter 7). Like Mark Zuckerberg of Facebook or Larry Page of Google, Bezos is a speed freak: as Brad Stone writes in his excellent history of Amazon, *The Everything Store*, 'he was constitutionally unwilling to watch Amazon succumb to any kind of institutional torpor, and he generated a nonstop flood of ideas on how to improve the experience of the website, make it more compelling for customers, and keep it one step ahead of rivals'.[79]

At the heart of the company's model is what it calls 'the virtuous flywheel'. As Amazon gets bigger, it is able to extract lower prices from its suppliers. It feeds these back to its customers, increasing its sales and size and enabling it to put even more pressure on its suppliers.

But Amazon, like other tech firms, is also ruthless about crushing the competition. Whenever it spots a challenge to its burgeoning monopoly in a particular product category, it is willing to do everything it can to acquire or destroy its rival – or, preferably, to put such pressure on that it sells out at a cut-rate price.

In the case of nappies, for example, Amazon threw $100 million down the drain in three months on its own bargain-basement delivery service in order to drive the existing market leader, Diapers.com, into its arms.[80] In its original business of books, meanwhile, the lawyers had to insist that the 'Gazelle Project' be renamed the 'Small Publisher Negotiation Program' – on the grounds that the idea of Amazon running these delicate, vulnerable creatures to the ground then clawing out their innards might not be the image the company wanted to project.[81]

Above all, however, Amazon's history shows the power of the platform – both as a spur to innovation in others, and as a bulwark against competition for its owner. Amazon's Marketplace, which permits third parties to sell goods on its site, now makes up 40 per cent of all its retail sales, shifting two billion units a year.[82] That's good for Jeff Bezos, and good for the sellers – but better for Bezos, given that Amazon's bots drive the prices of its own products to the lowest point possible, forcing others to match them.

As with Amazon's logistical services, in which it lets others use its warehouses, or its cloud computing business, the effect is to help others grow faster by carrying out vital functions more cheaply than they ever could – as long as they accept their place in Amazon's food chain. If I wanted to start a delivery business, for example, there is absolutely no way I could compete with the Amazon algorithms that compute thousands of alternative delivery mechanisms for every single package in order to minimise time and cost.

That is not just because Amazon is smart, but because it has invested hugely in solving these problems, in the process taking the average 'click to ship' time down from three days to four hours inside a year (the standard for the rest of the industry at the time was a comparatively sluggish 12 hours).[83] A further advantage is that Amazon is not just running these platforms, but using them as well: so if it sees rising sales in a particular sector in the Marketplace, or notes an upsurge in demand for a particular brand of T-shirt, it can position its own buyers to leap on the trend before anyone else even notices it.

The same is true for the other tech giants. YouTube has, under Google's stewardship, been a phenomenal success. Yet without Google's servers to keep the site going (at phenomenal cost), or its 'Content ID' algorithm to identify copyrighted content and stave off crippling lawsuits, YouTube would have found it far harder to sustain its explosive growth.[84] Bill Nguyen's Color was only one of many firms to try to challenge the big boys – and fail.

What truly makes these mega-firms different from their predecessors is that, for all their advantages, they constantly feel terrified of being displaced as the technological wheel turns. Indeed, the common thread between Mark Zuckerberg, Larry Page and Jeff Bezos – the owner-leaders of Facebook, Google and Amazon – is that they are not just messianic, but petrified. Petrified that something faster will come along and lure away their customers; that they won't or can't move fast enough to cope with the public's appetite for novelty.

Facebook's endless redesigns, for example, spring in part from its executives' (largely justified) conviction that they are the helpless slaves of its customers, the obedient instruments of their ever-greater craving for information. 'Mark's view,' David Kirkpatrick was told by Adam D'Angelo, one of Zuckerberg's oldest friends, 'is that Facebook had better not resist the trends of the world or else it'll become obsolete. Information is moving faster. That's just how the world is going to work in the future as a consequence of technology, regardless of what Facebook does.'[85] Bezos, too, once proclaimed that 'Amazon isn't happening to the book business. The future is happening to the book business.'[86] He is merely its instrument – and if not him, then someone else.

Their firms, therefore, put frenzied efforts into remaining agile and innovative. Despite their ever-increasing bulk, Facebook and Google have clung to their start-up mentality: keeping staff levels low and project teams tiny, and putting a premium on creative thinking. Both are also willing to disrupt themselves before others do it for them. When Google created a mobile-first messaging app called 'Inbox', the explicit goal was to blow its own Gmail out of the water; Facebook, virtually simultaneously, launched its own app called 'Rooms' to provide a faster and better version of its own core services. Larry Page has taken to using his mobile to do all his work, rather than a desktop computer: if smartphones are how billions of potential customers in Asia and Africa are accessing the web, he wants to make sure Google gets to them before anyone else does.[87]

Again, Reed Hastings is the poster child for this kind of behaviour – for reasons both good and bad. In 2011, Hastings announced – to general consternation – that the online portion of Netflix was being separated from the DVD-by-mail business (still the major profit engine), with the latter renamed 'Quickster'. But the move, allied to a hiking up of subscription fees, resulted in the loss of 800,000 customers; within a few weeks, Hastings was forced to announce that it had all been a very stupid idea, and things would go back to normal.[88]

What lay behind such an act of self-immolation? The idea, of which Christensen would approve, was to devote his and his best minds' full attention to the online arena, where the threat of competition and disruption was greatest. As Farhad Manjoo wrote, '[Netflix's] all-you-can-eat business model disrupted, and eventually killed, the previously dominant Blockbuster model for movie rentals. Hastings is likely paranoid, then, that Netflix is vulnerable to the same kind of disruption. And that's the logic behind the mail/streaming separation. Hastings would prefer to kill his own golden goose before anyone else beats him to it.'[89]

TODAY SILICON VALLEY, TOMORROW THE WORLD

The disruptive ethos may have been born in California's technology sector, but it has since spread to conquer the world. Indeed, this is the most obvious safeguard against any particular monopolist using its position to slow down the pace of change: the fact that such a monopoly will tend to be limited either to one particular area or to one particular iteration of the technology we use, and thus vulnerable to disruption from outside.

It is a truism, for example, to say that we now have a global marketplace, with more eager consumers and more dedicated entrepreneurs than at any point in human history. But that does not make it any less powerful a phenomenon. Today's executives know that they have customers – and rivals – everywhere from Shanghai to Saratoga to São Paulo. They also know that, in order to outcompete them, you need to move faster than them. The cumulative effect is to further accelerate the market.

There is also huge scope for this to increase: despite the tumultuous impact of globalisation, the world is still remarkably parochial, with the percentage of investment that crosses borders still in the single figures.[90] As faster communication and the pressure of competition erode national boundaries, new ideas will percolate back and forth ever more quickly – such as the dual SIM cards on knock-off Chinese phones that are now making their way into Western products, allowing people to switch seamlessly between work and home accounts. Asia, in particular, will provide a huge new market: by 2030, it will be home to 60 per cent of the global middle class.[91] But it will also prove a hotbed of innovation: China is already churning out more PhDs than America, and its own Googles and Facebooks and Amazons and Ubers are starting to stretch out their tendrils overseas.

In this chapter, we have seen how the great acceleration is being driven forward by a powerful alliance between hardware and software – between technology and ideology. Day by day, year by year, this process is reaching further into our lives, as the billionaires and would-be billionaires of Silicon Valley widen their ambitions. Space travel, education, energy, agriculture, transportation – all these sectors and more are in their sights. In their quest to disrupt the world, they and their cousins around the world are aided and abetted by the availability of raw computing power on a scale unimaginable a few decades ago. Yet the real reason this process of acceleration has such momentum, as we shall now see, is that it is exactly what our brains have programmed us to want.

2

Quick Reactions

'The actual substance of our daily lives is total distraction.'

– Jonathan Franzen[1]

There are many reasons we might not take George Miller Beard seriously today. The magnificently mutton-chopped psychiatrist from Montville, Connecticut, was a firm believer in the galvanising power of electricity, given to strapping his patients into an imposing apparatus, loosening or removing their clothes, and then passing a current of varying strength through them. He was also something of a bigot: he blamed many of America's troubles on 'the mental activity of women', claimed that democracy had had disastrous side effects (making 'every child, and every woman an expert in politics and theology', he wrote, 'is one of the costliest experiments with living human beings'), and believed that the 'lesser races' never suffered from tooth decay.[2]

But Beard – made famous by his investigation in 1878 of the 'Jumpers of Maine', a tribe of French-Canadian lumberjacks who possessed such a distorted startle reflex that they would obey any sudden command without thinking – has one significant claim to our attention. He realised that the cause of many of the symptoms he was seeing in his patients – including, but not limited to, 'insomnia, flushing, drowsiness, bad dreams, cerebral irritation, dilated pupils, pain, pressure and heaviness in the head, changes in the expression of the eye, neurasthenic asthenopia, noises in the ears, dribbling and incontinence of urine, falling away of the hair and beard, convulsive movements, a feeling of profound exhaustion, general and local itching' – was not physical, but mental.[3]

This new disease, Beard argued, was produced by civilisation itself. 'American nervousness, like American invention or agriculture, is at once peculiar and pre-eminent,' he wrote.[4] It was this pressure (and the United States' drier climate) that pushed his countrymen to speak more rapidly, to tune their musical instruments to a higher pitch, and ultimately exhausted what he called their 'nerve-force'. 'In every direction the modern brain is more heavily taxed than the ancient,' he wrote.

> The capacity of the nervous system for sustained work and worry has not increased in proportion to the demands for work and worry that are made upon it. Particularly, during the past quarter of a century, under the press and the stimulus of the telegraph and railway, the methods and incitements of brain-work have multiplied far in excess of average cerebral development . . . modern nervousness is the cry of the system struggling with its environment.[5]

Beard called this syndrome 'neurasthenia', but it soon acquired two more telling nicknames: first 'Americanitis' and then 'Newyorkitis'.[6] And his basic diagnosis – that the human system is struggling to cope with an ever more rapid environment – has never been more widely shared.

Today, the causes of stress and anxiety seem innumerable: robots taking our jobs, 'sexting' corrupting our children, the treadmill of work breaking our families apart, the internet turning our brains into flickering, hummingbird echoes of what they once were. We are becoming a society of shrinking time horizons, in which self-discipline and self-denial are giving way to instant gratification. That, or we are finding that the treadmill is moving too fast for us to cope: in her excellent book on the pressures of modern life, *Overwhelmed*, the *Washington Post* journalist Brigid Schulte cites surveys in which people claim to be too busy to go on holiday, to get lunch, to make friends, even to make love.[7]

Some commentators seriously suggest that, as a result of all this, we are doomed to a new dark age, or for humanity to divide into Eloi and Morlocks just as it did in H. G. Wells's *The Time Machine*. The neurologist Susan Greenfield, in her spectacularly didactic sci-fi novel *2121: A Tale from the Next Century*, envisages a formal separation between those who can tear themselves away from technology's teat and those consumed by what she calls 'the mindless squalor of Yakawow'.[8] 'The skills honed by video-gaming and information-processing,' she writes of her predicted future, 'gradually edged out those other human talents, such as understanding and wisdom.'[9] As a result, 'the raw feelings, the sensation of immediate experience' became all that mattered.[10]

But is this diagnosis correct? For all the doom-mongering, surveys show that most of us are basically happy, healthy and contented, and getting more so rather than less.[11] We all claim to feel under more pressure than ever, but that is the oldest tune in the hymnbook. 'Beyond doubt, the most salient characteristic of life in this latter portion of the 19th century is its speed,' wrote William Rathbone Greg in 1875. Victorians, he said, were living 'without leisure and without pause – a life of *haste* – above all a life of excitement, such as haste inevitably involves – a life filled so full . . . that we have no time to reflect where we have been and whither we intend to go'.[12]

This is the Ever-Waser argument all over again: technology changes, but human nature does not. If our attention spans are shrinking, why do we have so much time to watch hour upon hour of DVD box sets? Why, as we shall see later, is the length of the average *New York Times* bestseller increasing rather than decreasing? Why, amid the growth of messaging apps and txtspk, is the written word enjoying perhaps the greatest flourishing in human history?

This chapter will focus on the changes the great acceleration is making to our brains and bodies – which are so powerful precisely because, as with the services offered by the software companies in the previous chapter, they plug into some of our most basic biological cravings. In exploring these effects, there are certain limitations we must acknowledge. Much of the scientific evidence is mixed, or inconclusive. For example, we simply haven't been using the internet for long enough to produce proper, peer-reviewed studies of its long-term effects – indeed, the technology we use is changing so quickly that it can be near-impossible to reach a verdict on the costs and benefits of one particular form of it before we have moved on to the next.

That said, we already know enough to draw some pretty solid conclusions – in particular, about the effects of an accelerated lifestyle on the vitally important areas of attention, stress and sleep. Some of these changes may seem disconcerting, even alarming. Yet they are all, ultimately, within our ability to control. If we learn the right lessons, we will be able to lead not just faster lives, but better ones too.

BRAIN CANDY

Before we get to the consequences of the great acceleration, we need to start with the causes – to understand the neural mechanisms that make us crave speed and novelty. For it is precisely these mechanisms that are often invoked by those worried about the consequences of acceleration.

'Over the last few years,' wrote Nicholas Carr in the opening passage of his book *The Shallows*, 'I've had an uncomfortable sense that someone, or something, has been tinkering with my brain, remapping the neural circuitry, reprogramming the memory. My mind isn't going – so far as I can tell – but it's changing. I'm not thinking the way I used to think.'[13]

That's perfectly true – but we never were. Our neural circuitry is 'plastic': that is to say, it updates itself in light of experience, constantly forming new connections and pruning away old ones. This process happens in all of us – the very act of reading this text is rewiring your brain imperceptibly – but it takes place with explosive speed in children and teenagers. That is why we should worry far more about their exposure to potentially dangerous habits and technologies than about adults', since any damage will take much more effort to correct.

There are two other important properties of our brains that we need to understand before going any further. First, they are lazy (or thrifty, if you prefer), in that they constantly look for ways to save effort. Perform a particular action, and the pattern will be etched on your neurons. Perform it again, and it will be etched a little deeper. Pretty soon, the groove will be deep enough that the behaviour in question will have become a habit – handled by the 'lower' brain, without conscious thought, as part of an automatic and energy-saving process.

The second property is that our brains are junkies – constantly craving particular chemicals and sensations. In particular, when we experience pleasure, a part of the brain called the ventral tegmental area triggers the release of one of these chemicals, dopamine, and sends it to the appropriate regions of our brain. In his book *Pleasure*, the neuroscientist David J. Linden describes how activities as diverse as 'shopping, orgasm, learning, highly calorific foods, gambling, prayer, dancing till you drop and playing on the Internet' all evoke the same neural signals, which converge on a small group of interconnected brain areas called the medial forebrain pleasure circuit.[14]

The effects of this circuit are extraordinarily powerful. Linden describes how scientists in the 1970s created a 'Skinner box' (named for the psychologist B. F. Skinner) in which rats could push a lever to get a direct jolt of pleasure to the brain. The subjects would press the lever as many times as 7,000 an hour, ignoring their need for food, water or sex.[15] When similar techniques were tried on a human patient, 'he quickly began mashing the button like an eight-year-old playing Donkey Kong'. So powerful were the effects that scientists attempted to use this method to 'cure' homosexuality, first rewiring their patient's preferences and then bringing a prostitute into the laboratory in order to test the results.[16]

This pleasure circuit evolved, essentially, as a reward for good behaviour. The effect was to push us towards repeating the kinds of actions that made us more likely to survive. At its most basic, it means that when we like doing something, we want to do it again. The stronger the activation of the circuit, the greater the reward, the more deeply the pattern is etched on our brain – and the greater the potential for the act to become addictive.

With its endless stream of feedback mechanisms, the web – and much of our technology – could be precision-engineered to manipulate this pleasure circuit. Because of our pre-programmed craving for novelty (and status), getting a 'Like' on Facebook, receiving an email, or even just seeing the results come up for a Google search are each capable of triggering the circuit, with each one delivering a tiny high. 'Sites show you counts, totals, badges, because they know you'll come back to see them tick up,' says Christian Rudder, co-founder of the dating site OKCupid. 'Then they can put your increased engagement on a slide to impress their investors.'[17]

This plays into another of Skinner's discoveries: reinforcement schedules. To embed a particular pleasure response within a test subject's brain, you arrange a system in which a reward always follows a given signal. A green light goes on, a monkey gets a sugar drop; a phone goes ping, and you know you have a text message. Pretty soon, the activation of the brain's pleasure circuit is associated not with the reward itself, but the signal: not reading the text message, but simply knowing you've received it.

But Skinner also used his experiments with rats to demonstrate the power of variable schedules – altering either the time between rewards or the amount of effort it took to get them. He showed that even if rewards are random or uncertain, we still derive pleasure from the anticipation. Gamblers aren't excited just by the jackpot, but by pulling the lever or spinning the wheel. The same applies to checking our email – anticipation makes the heart beat faster, and if there's nothing worth reading the first time, we just know there will be the next.

The result is that, without realising it, we become trapped in Skinner boxes of our own making. As Carr says in *The Shallows*, 'If you were to set out to invent a medium that would rewire our mental circuits as quickly and thoroughly as possible, you would probably end up designing something that looks and works a lot like the internet.'[18] In a recent survey, people in 13 countries were asked what they would give up in exchange for internet access, if forced to choose. Overall, 75 per cent would give up alcohol, 27 per cent would forsake sex and 22 per cent would go without showers.[19] That proportion doubtless includes the US airline pilot and copilot who were so engrossed by their laptops that they flew 90 minutes past their destination (and forced Strategic Air Command to mobilise due to a suspected hijack).[20]

These effects embed themselves remarkably quickly. Studies have shown that asking non-web-surfers to spend just an hour a day online will rewire their brains to match the rest of ours within just five days.[21] This reward loop is, the psychologist Geoffrey Miller has suggested, a possible explanation for why we haven't met any aliens yet: they became so obsessed with triggering their pleasure reflexes that they didn't see the point of anything else, and promptly died out.[22]

Our basic biology, in short, programmes us to be fervent, even frenzied consumers of novelty and information (this is why the return journey on a trip to a strange place always seems much shorter – the brain is fascinated by new data, but sets aside the familiar). As the neuroscientist Daniel Levitin writes in *The Organized Mind*, 'humans will work just as hard to obtain a novel experience as we will to get a meal or a mate'.[23] And the acceleration of technology and society has also meant that there is more novelty, and more information, surrounding us than ever before – funnelled towards us through our phones, computers and email accounts.

ATTENTION DEFICIT?

Before getting into the possible consequences of this process on our daily lives, I should make one thing clear. Human beings are not rats in a Skinner box: just because you use Google, you are not doomed to end up hunched over a laptop at 2 a.m., oblivious to everything around you. Just because your children use Snapchat, they won't end up unable to read Proust.

And while our neural circuitry is indeed changing to fit the technology we use, this is not necessarily a destructive process. In a recent survey of 134 separate studies of the internet's impact on teenagers' brains, scientists at University College London concluded that there was no evidence that normal use – defined as under 30 hours a week – would cause any damage to cognitive ability.[24] Yes, there are conspicuous horror stories, but it is best to think of them as dangers to guard against rather than traps that are bound to swallow us.

For example, one of Carr's main concerns is that, by trusting Google to remember for us, we are destroying our memory. This would be doubly damaging if true, because memory is not just memory – it provides the cognitive parameters for our reasoning and understanding. To form new memories literally enlarges the brain, providing the space for new connections and new thoughts to be made. By some estimates, variations in this working memory account for 60 per cent of variations in IQ scores: the more you can remember, the smarter you are.[25]

Yet as Clive Thompson points out in his riposte to Carr, *Smarter Than You Think*, 'when it comes to knowledge we're interested in – anything that truly excites us and has meaning – we don't turn off our memory'.[26] We may rely on the internet to remember mundane facts for us, such as addresses or historical dates, but we still pay close attention to things that really interest us, such as the players on our favourite football team. It also depends on the type of online activity we're engaged in. Neural scanning of Twitter users by the company's London office has shown that even passive use of the service resulted in 34 per cent more neural activity in areas linked with memory formation than normal online use; with active Twitter use, the number rose to 56 per cent.[27] (Reading Twitter timelines, and tweeting, also resulted in much more activity in parts of the brain connected to emotion.)

When it comes to attention, however, there is certainly cause for concern. In Carr's phrase, the internet 'seizes our attention only to scatter it'.[28] Email and web-surfing are technologies of instant gratification, delivering distraction and information on demand. In large part, this is due to the sheer volume of information. As John Freeman notes in *The Tyranny of Email*, we check email 30 or 40 times an hour; we ruin our backs and our eyes (in Singapore, Freeman points out, 80 per cent of children are myopic, up from 25 per cent just 30 years ago).[29] Among smartphone owners, 79 per cent claim that checking their phones is the very first thing they do in the morning, even before getting out of bed.[30] We check them on holiday, during meetings, in the supermarket queue, even on the loo.[31] In one US study – carried out in 2003, well before smartphones and apps and status updates dug their claws into us – only three out of 220 university students were able to turn off their mobile phones for 72 hours.[32] In another, cutting people's desktop internet connection caused them to become snappy and agitated, to the point where 10 per cent admitted to physically assaulting the computer.[33] Even the knowledge that we have an unread message in our inbox can diminish our effective IQ by 10 points.[34]

At the most basic level, this bombardment of email and other messages deprives us of time that could be more productively used. By some estimates, reading and processing emails hogs between 30 and 50 per cent of the working day.[35] Surveys cited by Brigid Schulte, whose work I have drawn on extensively in this chapter, found that two-thirds of workers feel they don't have enough time, and 94 per cent have at some point felt 'overwhelmed by information to the point of incapacitation'.[36] It was actually in part to solve this problem that Apple came up with its Apple Watch – the intention was that it would filter the endless nagging notifications from your iPhone, which even Apple admitted had assumed too great a dominance over users' lives.[37]

This bombardment of information also means that we are plagued by interruption – all those little pings and notifications that we are biologically programmed to pay attention to. Research by the sociologist Gloria Mark found that the average employee manages to spend just 11 minutes on any given task before switching tack, and changes focus within a given task every three minutes; much if not all of this is driven by messages popping up on screen.[38] So chopped up is the workday that Fortune 500 CEOs average, by one estimate, just 28 uninterrupted productive minutes a day.[39]

This herky-jerky lifestyle has hugely important consequences. First, in the words of the psychologist Herbert Simon, 'a wealth of information creates a poverty of attention'.[40] We are so distracted that the art of concentration becomes increasingly hard. Teachers, lecturers and parents complain of children's inability to knuckle down, to grapple with complex ideas. Attention spans are shortening: advertisers who used to prepare five-minute videos to pitch their services to potential clients have cut them down to 90 seconds.[41]

Many people now worry that, as our brains grow used to a diet of empty calories, they will be less able to digest more nutritious fare, such as books or even complex ideas in general. In one tantalising but limited study, Chinese researchers found that in internet addicts, the areas of the brain that relate to attention, control and executive function bulk up with 'white matter' – extra nerve cells built for speed – rather than the neuron-rich 'grey matter' that is better for cogitation.[42] Moreover, it found a 'shrinkage of 10 to 20 per cent in the area of the brain responsible for processing of speech, memory, motor control, emotion, sensory and other information'.[43]

It's not just internet addicts who are affected. As Maggie Jackson warns in her book *Distracted*, 'studies show that many US high-school students can't synthesise or assess information, express complex thoughts, or analyse arguments'.[44] Maryanne Wolf, an expert on the science of reading from Tufts University, threw her weight behind a campaign for 'Slow Reading' (in imitation of the Italian-born 'Slow Food' movement) upon finding herself disgusted by her inability to get through a Hermann Hesse novel after a few years of internet surfing.[45]

Part of the problem is that, in a perfect example of the great acceleration reinforcing itself, the technologies that push us further and faster also render us less able to cope with their effects. Interruption and distraction, for example, change from things we put up with to things we crave, because of those little dopamine hits they deliver. Soon, even the briefest gaps in our schedule demand to be filled – usually with little snacks on our mobile phones. In one study, researchers asked people to spend just 15 minutes alone in a room with their thoughts. More than half confessed to not enjoying the experience. So unpleasant was it, indeed, that in a follow-up experiment, many chose to experience an unpleasant electric shock to relieve the tedium (a whole other book could be written about the fact that two-thirds of the male subjects took this option, but just a quarter of the women).[46]

Some people have attempted to defend this kind of hummingbird mentality as an example of productive multitasking: for example, arguing that the teenager flicking between instant messaging, playing a computer game and doing their homework is acting more like a chef 'keeping pots cooking on a hob, checking each one as and when'.[47] By bouncing between different projects and ideas, we may feel we are being supremely productive.

But we are not. Multitasking is, as experiment after experiment shows, a myth. 'What looks like multitasking is really switching back and forth between multiple tasks, which reduces productivity and increases mistakes by up to 50 per cent,' argues the author Susan Cain.[48] Levitin is even more blunt: when we multitask, 'we unknowingly enter an addiction loop as the brain's novelty centres become rewarded for processing shiny new stimuli'.[49]

This shortening of attention spans has other consequences, too. One of the most important is that it also curtails our ability to delay gratification. 'The most far-reaching social development of the early 21st century,' argues my former colleague Damian Thompson in his book *The Fix*, 'is our increasingly insistent habit of rewarding ourselves whenever we feel the need to lift our moods.'[50] He sees addiction spreading across society in ways large or small – not least in the consumption of mid-afternoon cupcakes as a 'treat' for getting through the bulk of the day.

Some would argue that the way that our brains flit from web page to web page, or that our hands reach out and grab whatever we want, signals a failure of willpower. In that, they are more right than they know. Paying attention takes not just mental but physical energy – in the form of the glucose that feeds mental activity. And as Daniel Goleman says in his book *Focus*, 'the signs of mental fatigue, such as a drop in effectiveness and a rise in distractedness and irritability, signify that the mental effort needed to sustain focus has depleted the glucose that feeds neural energy'.[51]

Even when we have trained ourselves to concentrate, that capacity – known by experts as executive attention – is eroded by stress or overwork or fatigue. 'Life immersed in digital distractions,' says Goleman, 'creates a near-constant cognitive overload. And that overload wears out self-control.'[52] This is one reason why stress is frequently accompanied by weight gain: it is when we are on autopilot that we reach for the chocolate bar, or the full-fat latte, to keep us going.

The result is a double trap. We are less likely to focus, because our minds are being trained by technology to move as fast as it does. And we are less able to focus, because the environment that the same technology fosters has weakened our energy reserves. Daniel Kahneman, in his book *Thinking, Fast and Slow*, describes this state as being 'cognitively busy': people in such a condition, he says, are 'more likely to make selfish choices, use sexist language and make superficial judgments in social situations'.[53] He cites the case of a parole board in Israel, whose members granted 65 per cent of requests for furlough after a decent lunch, and almost none before it: when their mental batteries were drained after a long morning's work, they defaulted to the simplest, easiest option.[54]

And this raises one of the most critical points of all: the way that the great acceleration's effects are often felt most strongly by those who are least able to cope. A member of the professional classes may be worn out after a series of 12-hour days at work, or coping with screaming children, but they probably have the cognitive resources to just about cope. For those whose life is a day-to-day struggle, however, it becomes almost impossible to regain the focus and control needed to plan for the long term.

In their book *Scarcity*, Sendhil Mullainathan and Eldar Shafir describe how the experience of poverty in particular reduces our mental bandwidth: it saps our executive attention and cognitive control, and induces a state of 'tunnelling' in which only the most immediate problems can be focused on.[55] It is when we tunnel, they argue, that we take out the payday loan, or reach for the cupcake: it shortens our time horizons to the point that we ignore the future costs.

This isn't just a trap for the poor, however. Information overload – and the pace of life in general – rob us of the time to stop and think. For example, brain scans of chief executives show that a form of 'tunnelling' takes place when they are relentlessly driving organisations forward, so that it takes a huge cognitive wrench to stop single-mindedly exploiting the current market niche or business model and start to think more broadly.[56]

This information bombardment doesn't just inhibit our ability to think deeply or broadly – it narrows our bandwidth too. This is why, notoriously, so many inspirations come to people in the shower or the bath: it is when we relax and allow our mind to wander that the alpha waves are generated that promote what can, for want of a better term, be called creativity. In what is known as the 'novel uses task' – in which a scientist asks their subject to think of all the possible uses for a brick or other object – those who let their mind roam come up with 40 per cent more ideas.[57]

The toxic combination of more information and less time to process it is another of the feedback loops with which the great acceleration is rife. And it turns out that it is not merely preventing us having more creative thoughts: it may also be turning us into worse people. As mentioned earlier, when Philip Zimbardo asked *USA Today* readers in 2008 to report their level of annoyance with day-to-day nuisances such as queuing or waiting for deliveries, he found that their collective frustration and annoyance had soared since it carried out a similar exercise in 1987 (hence Google's obsession with loading times).[58]

Similarly, Alan Johnston, a Westerner held hostage in Lebanon, recalled how his flight back from captivity was delayed by an hour because someone had taken their chihuahua through security. 'Everybody was so annoyed that it was slightly surreal. I couldn't work out why they couldn't bear waiting for just for one hour. Yet within six weeks of getting back to London I remember waiting at a bus stop and swearing because there weren't any buses coming along. Already the old impatience had come back.'[59]

The more time pressure we are under, the more impatient we get – and the nastier. In 1977, students at Princeton Seminary were asked to give a presentation in a distant room; some were told they had plenty of time to get there, others that they were already late. On the way there, they passed by someone slumped and coughing in an alley. The majority of students under no time pressure stopped to help; more than 90 per cent of those who believed they were late simply hurried on by.[60]

If time pressure does indeed erode compassion, then the quickening tempo of life may even be another explanation for one of the social trends that has most alarmed the critics of acceleration – the growth of narcissism, and a perceived collapse in empathy. Surveys of American teenagers, for example, have shown a monumental rise in self-obsession.[61] The suggestion is also that such children are less psychologically robust than the previous generation.[62] When everyone spends their days tending their social media profile, is it any wonder that we are breeding people who are not just selfish but somehow fragile – 'screenagers' who expect life to come with an undo key?

STRESSED OUT

The internet's impact on our attention span is certainly a significant issue. The real problem, however, is that is does not operate in isolation. It is not just that we are deluged with information, with requests and messages; we are also told that our self-worth and happiness depend on how many of those demands we can answer.

According to a recent YouGov survey, Britons find their jobs more stressful, precarious and demanding than ever before.[63] In a wider survey of 115,000 people across 33 countries, 13 per cent said they had trouble sleeping due to work, and 40 per cent said taking sick days made them feel guilty.[64]

This is what Brigid Schulte describes as 'The Overwhelm' – a state of mind in which we are constantly scattered, fragmented and exhausted, in which we feel that no matter how hard we race, how long we work, we are falling short of the mark.[65]

The fundamental problem here, as with our reward circuits, is that a stress system designed for Paleolithic times is a poor match for 21st-century life. When we spot a threat, our amygdala sends out the familiar fight-or-flight signals. Hormones flood our body: adrenalin gets the heart pumping faster; the brain seethes with dopamine, shutting down our higher reasoning and letting our basic instincts take hold; cortisol ramps up the production of sugary glucose, giving our muscles the boost we need.

Such experiences are not only natural, but necessary. Like athletes, we are designed for these sudden bursts of activity. They may even be fun: many of the hormones involved have a strong effect on the brain's pleasure circuit, making us feel powerful, effective and in control (which is why some of us are literally addicted to stress: it gives us the same chemical hit as drugs or alcohol).

Yet stress today is not a rarity, but a constant. Instead of physical threats, we face a constant bombardment of external information – information designed to be as attention-grabbing and (in the case of TV news) fear-inducing as possible. Lights, sounds, shocks – these are the kind of things that evolution has hard-wired us to pay attention to, to ensure our survival, and we confront them minute after minute.

Making things worse is the fact that, while the initial fight-or-flight reaction may be instant, the stress–response system takes a long time to calm down once triggered. As the scientist Robert

Sapolsky has written, 'the autonomic parts of your body move like a freight train: they build speed gradually and take a long time to come to a stop'.[66] This creates its own feedback loop: when something makes us worried or angry, our bodies still remain in an agitated state even when our minds have cleared. That in turn helps to induce the emotional state again: our brains pick up the signals from our bodies and decide there must be a reason for them.

When we are under severe strain, or of a particularly nervous predisposition, the cortisol and other hormones remain in our system, and stress gradually becomes less an exceptional experience and more a way of life. As Sapolsky explains:

> When an organism is confronted with some sort of threat, it typically becomes vigilant, searches to gain information about the nature of the threat, struggles to find an effective coping response. And once a signal indicates safety – the lion has been evaded, the traffic cop buys the explanation and doesn't issue a ticket – the organism can relax. But this is not what occurs in an anxious individual. Instead, there is a frantic skittering among coping responses – abruptly shifting from one to another without checking whether anything has worked, an agitated attempt to cover all the bases and attempt a variety of responses simultaneously. Or there is an inability to detect when the safety signal occurs, and the restless vigilance keeps going.[67]

The chaotic and unpredictable nature of modern life does not help matters either. Just as we are programmed to like random rewards, we are programmed to dread unpredictable dangers and punishments. Joan Silk, a primatologist at UCLA, has shown that alpha-male baboons keep their packs cowed in large part via sudden and unpredictable acts of brutal aggression (behaviour you may well recognise from your own workplace).[68]

Cortisol is not actually the cause of stress; in fact, its function is more to return our systems to homeostasis, like the ambulance that is always seen at a crash site. But when our systems are flooded with cortisol, kept constantly under stress, extremely bad things happen. Stress, says Schulte, 'weakens the body's immune system, making it more susceptible to inflammation, cardiovascular disease, high blood pressure, type 2 diabetes, arthritis, osteoporosis, obesity, Alzheimer's disease and other debilitating ailments' – including chronic fatigue syndrome and cancer, as well as depression and anxiety, especially in women.[69] Within our cells, the telomeres – the chromosomal equivalents of the aglets on shoelaces, which keep our DNA from fraying – are shorter in those who experience significant stress, which can lead to wrinkled skin, greying hair, sagging muscles, impaired eyesight and hearing and lower life expectancy.[70] Elevated stress levels have even been suggested as part of the reason why birth rates have fallen in the West – and why more girls are being born than boys.[71]

As if all this weren't enough, stress also weakens the brain, by breaking down the neural loops that hold short-term memories together as they are transferred into long-term storage, and shrinking the prefrontal cortex, the part of the brain that governs our capacity for focus and attention – another example of feedback loops at work.[72] It was previously thought that this only applied in extreme cases, such as PTSD or extreme childhood trauma. But that is not so. Schulte profiled Emily Ansell at Yale, whose research has found that the greater the amount of stress we experience, the smaller the volume of grey matter in our brains. In a study in 2012 of more than 100 healthy subjects, those who had recently experienced a stressful life event, such as a job loss or divorce, 'showed markedly lower gray matter in portions of the medial prefrontal cortex, an area of the brain that regulates not only emotions and self-control, but physiological functions such as blood pressure and glucose levels'.[73]

As with all of the costs and benefits of the great acceleration, the impact of stress is not evenly distributed. For example, those who are most vulnerable will tend to suffer most, and find it hardest to bounce back. Ansell found that those who have had stressful experiences in the past suffer proportionately greater grey matter loss from further stress – and are, as other studies show, less able to repair the damage.[74]

There is also something of a genetic lottery at work. It turns out that the enzyme that clears dopamine out of the prefrontal cortex – COMT – comes in two varieties, weak and strong. In those of European descent, a quarter of us have only the weak version of the enzyme, and a quarter have only the strong, which is four times more powerful.[75] Those with only the weak version (dubbed 'Worriers' by Po Bronson and Ashley Merryman in their book *Top Dog*) get overwhelmed by stress, whereas those with only the strong version ('Warriors') thrive on it, to the point where they are actively hampered by the absence of dopamine in unstressed circumstances.[76]

These effects shouldn't be overstated – studies show that Worriers can cope just as well, if not better, with stressful situations, such as piloting a plane, if they have experience and training[77] – but they are nevertheless significant. And crucially, the presence of oestrogen alters the transcription of COMT so as to slow down dopamine reabsorption by 30 per cent.[78] This means that the cliché about women being worse at handling stress is grounded in fact: whereas men under stress tend to tune out emotional cues and make more rational decisions, women will look for and follow emotional cues to calm themselves down, and so – in the broadest of generalisations – need to be supported rather than shouted at. (If it's any consolation, women make much better financial analysts, because they are able to leave their ego at the door and avoid making risky bets.)[79]

So we all have different stress thresholds, and experience it in different ways – often due as much to cultural conditioning as to biology. In the West, for example, highly stressed people often get crippling lower back pain. In Japan, they become convinced that their testicles have withdrawn into their bodies; in Korea, that they are in burning agony; in China, that they have a shoulder or stomach ache.[80]

Yet whatever our personal mode of suffering stress, its impact on our well-being is enormous. Researchers have found that the single best predictor of people's health is how they feel about the level of stress in their lives.[81] Which is a problem, because the modern lifestyle could be precision-engineered to increase it.

CHAINED TO THE DESK

Long hours are, of course, a feature of workplaces in many countries. But as with much else in the great acceleration, the pattern for the rest of the world was set in the US. As Schulte says, nearly 40 per cent of American men with a degree work more than 50 hours a week – as do 32 per cent of professional single mothers.[82] In 2014, just 56 per cent of Americans took even a one-week holiday, as against 80 per cent in 1976 (and pretty much 100 per cent of present-day Britons and Europeans).[83] She quotes a travel expert describing how the trend for week-long backpacking trips has given way to weekend getaways, then to day trips. 'By 2010 ... the time focus narrowed to what people could do in about four hours in an afternoon ... and now, the focus is on what you can do in 45 minutes over lunch.'[84]

In the finance and technology sectors, which live or die by acceleration, the trends are even more extreme. In Silicon Valley, those putting in less than 90-hour weeks – those who don't live in a state of permanent crunch, scarfing down snacks at their desks as they kill themselves to get the product out – are widely seen as losers. The pressure comes not just from above, but from your colleagues: for an extreme example, see the brutal working conditions uncovered by a *New York Times* investigation into Amazon, which it accused of 'conducting a little-known experiment in how far it can push white-collar workers' (one woman whose child was stillborn was allegedly told to make sure her focus remained on her work).[85]

The situation in Wall Street is equally intense. A study in *Sociological Quarterly* reported how bankers who had escaped to other fields after the financial crisis had exported their long-hours culture: 'I have made a comfortable life for myself here,' said one new arrival in Arizona. 'There is hardly a day when I have to be in the office later than 11 p.m.'[86]

Long hours don't necessarily have to equal unhappiness: many of us love our jobs and thrive on the challenges they present. In one outstanding example, an insurance firm called Amerco became worried by the hours its staffers were putting in, and by their complaints about the strain this was putting on their family life. It devoted huge resources to promoting flexible working practices, with promises that no one would be punished for doing so. Yet only 4 per cent of workers with children cut back, and only 1 per cent started working from home. The problem was that the firm had already done too good a job: spending time at the office, working on important and interesting projects in a pleasant environment, was far more appealing than going home to argue about whose turn it was to wash the dishes.[87]

But for many people, long hours are a curse – and a health hazard. 'People think of the working day as starting when you're at your desk,' says a City analyst who asked me to refer to him as Simon.

> But that's not how it works any more. It starts the moment you wake up in the morning – you switch on CNBC, you pick up your BlackBerry to check on Bloomberg, to see what's going on, and what it means for you and your company. Your brain's in gear from six, six thirty – and then it's into breakfast meetings, research,

> one-on-ones with clients, briefings, flights. Even when you're on the train home, or eating dinner with your family, the phone will go, and you'll have to talk to the office in New York, and see where things have finished for the day. At some point, you find that you just can't get out of bed in the morning.[88]

A couple of years ago, Simon was diagnosed with chronic fatigue syndrome: when he spoke to the specialists, he was told that 95 per cent of the cases they were seeing involved City workers just like him. Work-induced stress is, he explains, the chief suspect: 'In the old days, if you confronted a wild animal in the bush, you'd deal with it and calm down again. In the modern markets, your fight-or-flight mechanism is permanently switched on. Sooner or later, it does so much damage that your body basically shuts down to recover.'

A massive survey by the Canadian health service, again cited by Schulte, found exactly this effect. 'As work weeks get longer and leisure time shrinks, people are becoming sicker, more distracted, absent, unproductive and less innovative,' it said.[89] It concluded that: 'The link between hours in work and role overload, burn-out and physical and mental health problems [suggests] that these workloads are not sustainable over the long term.' The latest research has found that working more than 55 hours a week raises your odds of a stroke by a third, in addition to raising the odds of a heart attack.[90] And such problems, it should be noted, become contagious: researchers in Germany have found that being around those under stress (or even seeing them on the TV screen) makes us stressed ourselves.[91]

These extra hours aren't just unhealthy – they're actively counterproductive. In an in-house survey in 2005, Microsoft employees reported that despite the company's long-hours culture, they only actually scraped together 28 productive hours a week (due in part to all the interruptions mentioned above).[92] Research undertaken in the Eighties found that you could get short-term gains by pushing people to 60 or 70 hours, but only for a couple of weeks – and other studies have found that overwork leads to errors that take longer to fix than the extra hours worked.[93,94] Schulte cites an experiment that compared two groups of workers at a Boston consulting firm: some worked 50 hours plus, didn't use their holiday time, and were tethered to their office by their phones. The others worked for 40 hours, took their vacations, switched off their phones. This group were not just more productive, but hugely so.[95]

Entrenching a commitment to work-life balance is, as we shall shortly see, one of the main ways in which we can turn the effects of the great acceleration to our own advantage – to hack the system so that it delivers what we want. Yet in the increasingly chaotic world that acceleration brings, it may grow ever harder to do so: when people are more frightened of losing their job to foreign competition or to automation (the great acceleration at work again), they will be more inclined to heed their bosses' calls for greater dedication.

The truth is that the more we work – and the more we allow it to invade our leisure time in the form of messages and emails – the more squeezed the rest of our lives become, and the higher our stress levels. In particular, while 'having it all' has become something of a cliché, we still cannot stop ourselves trying to be perfect workers, perfect partners and perfect parents, all at the same time.

For example, even though far more women have jobs than in the Sixties, the average mother now spends more time actively looking after her child.[96] The amount of high-quality 'interactive care' – reading to and playing with your children – has tripled.[97] But we still feel it isn't enough. The result is a devotion to parenting that borders on obsession, certainly according to the hours that are being put in.

This shows again how aspects of the great acceleration feed off and reinforce each other. It also shows how it acts to polarise society – in this case between a globalised professional class, for whom there can never be enough hours in the day, and those on the fringes of the labour market, for whom there is never enough work to go round.

This has malign effects at both ends of the spectrum. At the top end, it puts intense pressure on professional couples to spend far more time with their children. In a fast-paced, technology-driven world, having the skills to understand and manipulate that technology becomes increasingly important. As the world gets richer, it has fewer children, due in large part – in developed countries – to more women being in work, and that work placing more and more pressures on them.

This means that parents are now putting all their eggs in one or two baskets – so if good exam results have become the golden ticket to a good life, they will do everything they can to make sure that their children get them. This, in turn, means hothousing and helicopter parenting to the point of mania – even though those same parents are already more than occupied with the increasing demands of the workplace. A revealing study of air traffic controllers found that the quality of their parenting exactly matched the number of planes they had had to deal with on that particular day: the more they had to focus on work, the less they wanted to engage with their children on returning home.[98]

In recent decades, the time children have for unstructured play has been sliced to the bone – which perhaps explains why, according to one eye-opening study, the children of the affluent are two to three times more likely to suffer from depression or anxiety than those raised in harsh urban poverty, and more likely to use drugs and alcohol.[99] It is also why, as we will see, so many of these smothered, over-pressured children retreat online for a respite – prompting further worry from their parents.

Yet things are no better for those in more menial careers. One of the rewards of being a high-flyer is feeling that you have control of your own destiny, or at least superior status, alleviates or even eliminates the effects of stress: studies of the British civil service have shown that bosses live longer, healthier lives than their underlings.[100]

Being a grunt is not only bad for your health, and your lifespan. In a mirror of the research we discussed earlier about 'tunnelling', it also shrinks your time perception: the less power you have, the more pressured and time-poor you feel. Perhaps the best example comes from a study of sugarcane farmers, which measured their intelligence before and after the money from the harvest had come in. It found that, when the cash ran out, the farmers dropped 9 or 10 IQ points – not because they had become more stupid, but because their economic predicament swallowed much of their mental capacity.[101]

Stress, in short, is the new pandemic – its corrosive effects felt in different ways in different places, but always pushing us to react to the great acceleration not with excitement, but with fear and alarm. And the quicker the pace of life gets, the more heavily we feel its effects.

SLEEPLESS NIGHTS

The human body does have a wonderful recovery mechanism – sleep – whose purpose is to cope with stress and strain. Yet this too is being eroded by the pressures of acceleration.

There are various explanations for why we sleep. Partly, it is about power-saving: it shuts down parts of our energy-intensive brain so that we can restore and recuperate. Partly, it is about information-processing, as we rehearse and reconfigure problems from the day and see them in a new light. It also prunes the synapses formed by all those conversations and encounters during the day, rewiring the neural rewiring so that the brain does not become overcrowded. This is why a bad night's sleep can blur your memory of an event, even two or three days afterwards (and why drinking is associated with memory loss, because it disrupts this system).

What is crystal clear, however, is that sleep is essential, and that both its quality and quantity are in decline. In 1960, most Americans slept for between eight and nine hours a night. By the turn of the millennium, that had fallen to under seven, with one in three people getting less than six (the threshold at which sleep loss starts to affect our day-to-day performance).[102] In a recent survey of 38,700 British employees, only 15 per cent felt refreshed by their sleep.[103] 'We are, as a population, sleeping less now than we ever have,' wrote Maria Konnikova recently in the *New Yorker*.[104]

The problem isn't simply the number of hours we're sleeping for: it's much more interesting than that. It turns out that we actually have two sleep systems, rather than one. The first is the simple winding up and down of the clock that we call tiredness – the longer we are awake for, the more we need to sleep.

The second system is more complicated. This is the biological clock – the circadian rhythm – which tells us when it is time to go to sleep. It is set, we now know, not by one master circuit in the brain, but by billions of tiny clocks in every cell of your body. The reason why we get jet lag is that these separate clocks get out of synch, as our alertness or digestion or light–dark perception drift out of kilter with each other.[105]

This is also, some researchers believe, one of the reasons for our rising levels of sleep disruption: parts of this clock are set by when we eat and the amount of sugar in our bloodstream, and those parts are sent haywire by the inter-meal and late-night snacking that we are increasingly prone to. This does not just affect our sleep, but our health: these clocks control processes such as digestion, detoxification, DNA damage repair, circulation, cholesterol and many others that are directly linked to illness and disease.[106]

The great acceleration, it turns out, has nudged our bodies out of synch with the day/night cycle – and we are suffering for it. Perhaps the most influential modern sleep researcher, Till Roenneberg, has described this phenomenon as 'social jet lag': the fact that we are all entrained by a schedule set by work and not by the solar cycle.[107] When we catch up with our sleep at weekends, we are partly paying down the 'sleep debt' caused by not getting enough shut-eye during the working week. But we are also, in effect, travelling back from the 'office' time zone to our natural schedule.

This is self-evidently a bad thing – especially when the dislocation is more severe, as with night work or other antisocial schedules. Researchers have taken genetically identical populations of mice and subjected some of them to the equivalent of shift work. 'In most studies,' explains Steve Brown of the University of Zurich, 'these mice die earlier, they have immune problems, they have digestive problems, they have increased cancer, they have increased cardiac problems – you can go through the whole range of things you wouldn't want to have.'[108] Studies of human workers, such as nurses in the United States who have to be on call overnight, have shown similar effects, to the point that shift work has actually been classified as a probable carcinogen by the World Health Organisation.[109]

Brown cites experiments in which scientists monitored junior doctors – a peculiarly sleep-deprived group – while on call, and even persuaded them to wear brain monitors. What was found was that as they became sleepier, 'micro-sleep' events were happening all over their brains, as particular bunches of neurons shut down. 'Doctors would either fully or partially go to sleep in the middle of an operation,' explains Brown. 'Even doctors talking to patients would have a large portion of their brain showing slow-wave sleep patterns.'[110]

This kind of sleep deprivation has been the culprit in all kinds of accidents, ranging from Chernobyl to the *Exxon Valdez*.[111] Lack of sleep is as damaging to cognition as being drunk, or more so – which is why many sleep scientists will not take taxis after midnight, for fear of 'catastrophic naps'.[112] And Roenneberg's research using students as guinea pigs has suggested that the effects are not limited just to those on the night shift. True shift work, he argues, is only an extreme form of circadian misalignment: he claims 'the majority of the population in the industrialised world suffers from a similarly "forced synchrony"', with associated costs to its health and well-being.[113]

It is not just social jet lag that is causing problems. 'There are so many areas now where we have good data suggesting that over-riding internal biology has a big effect on our sleep, and therefore our health,' says Russell Foster of Oxford University.[114] For example, sleep deprivation in children causes hyperactivity – frequently leading to mistaken diagnoses of ADHD.[115] One connection that researchers are interested in pursuing – although it may take many years to demonstrate it – is between sleep deprivation and Alzheimer's. The reason, as Steve Brown explains, is that sleep acts to clean out the macro-molecular junk from the brain, by dilating the lymphatic vessels. These include the deposits that form the plaques that make up the substrates of Alzheimer's disease. So there is a prima facie case for suggesting that chronic sleep deprivation could be a risk factor.

The more immediate problem comes when lack of sleep is combined with a more and more stressful lifestyle. To get going in the morning, and keep going throughout the day, you need to rely on coffee, or Coke, or high-calorie snacks. That leaves you too wired to sleep, so you resort to sedatives such as sleeping tablets. The problem is that these don't actually increase the amount of decent sleep you enjoy – so when you get up the next day, you need more stimulants to set you going. This is what Foster calls a 'stimulant–sedative loop': it has, he says, long been seen in shift workers, but now it's found at all levels of society. He explains:

> I spoke to a 13-year-old kid at a school in Liverpool and asked: 'What's your sleep like?' And she said: 'It's fantastic, it's great.' And I

said: 'Wonderful, so what's your tip?' And she said: 'Oh, well, I take my mother's sleeping tablets.' This is a 13-year-old child using sedatives. So I asked her: 'How do you feel the next morning?' And she said: 'Well, pretty groggy. But I'm OK because I've had about three Red Bulls by lunchtime.' This is a massive stimulant–sedative insult to a developing, plastic brain. And it arises because we're trying to squeeze more and more into an already overcrowded day, and the first thing to go is always sleep.

The effects of sleeplessness, of the 24-hour lifestyle, are certainly bad for adults. Says Foster: 'If you're in this stimulant–sedative loop, because booze is cheap, you slide into a state of sedation to counteract the stresses and the tiredness and the stimulants you've been taking. You're not fit as a partner, let alone as a parent.' Lack of sleep also inhibits creativity – an ever more vital resource in the knowledge economy.

As with so many of the effects described in this chapter, the problems of sleep and stress feed off each other. If we're sleep-deprived, we can override our bodies' need for rest by activating our stress responses, and feeding more glucose into the system. This is what shift workers do – which explains why they suffer from higher rates of cardiovascular disease, from pushing this mechanism too hard.[116] Their elevated levels of cortisol, meanwhile, depress the immune system and result in a higher rate of infections and health problems. And, crucially, when we are stressed we find it harder to get to sleep – pushing us into a downward spiral of greater tension and less rest.

Just like the internet's rewiring of our brains, the impact of sleep deprivation is rapid (and insidious, given how stunningly bad we are at realising how affected we are by it).[117] In one experiment, Eve Van Cauter of the University of Chicago took two groups of healthy young males in their early twenties.[118] One was only allowed four hours of sleep a night, and the other up to ten. After only seven days, levels of the hunger hormone ghrelin had soared in the sleep-deprived group; carbohydrate and sugar consumption had shot up and the ability to clear glucose from the bloodstream was borderline-diabetic. Without pressing the point, Van Cauter notes in the study that the rise in obesity in recent decades across society as a whole precisely mirrors the fall in sleep duration and quality.

As with the internet's impact on the brain, the effect of poor sleep on children and teenagers should cause even more concern. The veteran sleep scientist Jim Horne is a strong critic of the argument that we are becoming an increasingly sleepless society (even though most of his fellow researchers disagree). But even he agrees that children are spending too much time staring at screens when they should be settling down for bed. This goes for adolescents too, who are biologically programmed to need more shut-eye than adults.[119] Yet the average US teenager, a study in 2010 found, sends 34 texts a night after their bedtime.[120]

The problem, says Russell Foster, 'is that bedrooms now for teenagers are places of entertainment, not sleep. So a predisposition to go to bed late and get up late is enormously exaggerated by social media and other electronic devices.'[121]

Parents' reluctance to provide strict guidelines also plays a part. According to Foster:

> We feel uncomfortable about providing guidelines about sleep to our young people, and they're almost crying out for it. Mood fluctuations, depression, anxiety, frustration, anger, impulsivity are all hugely augmented by lack of sleep – you speak to these incredibly dedicated teachers and they talk about kids falling asleep for the first few hours when they come into school. This marginalisation of sleep I think is having a major impact on our health and our quality of life.

THE ROAD TO RECOVERY

The biological effects of the great acceleration are certainly profound – and for many, profoundly damaging. Yet paradoxically, they also offer us cause to be optimistic. Many of the problems currently afflicting us have come about through ignorance: we did not realise what information overload was going to do to us, even though we dimly felt its effects. But there is an increasing body of evidence showing that all of these problems can be addressed – that there are relatively quick and easy fixes that can position us to enjoy acceleration's benefits without succumbing to its drawbacks.

Take sleep loss. In large part, our sleepless nights are because we insist on staring at screens before going to bed – and the frequencies of light involved prolong our wakefulness.[122] But we don't even need to set aside our phones, just to use filters on the screens (or special sunglasses) that convert the blue light to orange, which is much more neurally soothing.[123]

In the more distant future, a technology called transcranial magnetic stimulation – essentially, a mobile phone on steroids – could induce neurons into a state of slow-wave sleep: imagine pop-up sleep booths which would help us to free up even more of our day. But even now shift workers could easily be tested for their 'chronotype' – whether their genetics predetermine them to be a morning or evening person. This could be accompanied by extensive education as to the dangers of the work, the provision of proper nutrition, and regular tests of stress levels.

As for those Red Bull-swigging teenagers, it would be a relatively simple matter to add to the curriculum classes explaining the importance of proper sleep, and to teach parents about it. Something as simple as setting family rules for the consumption of media cuts the time children spend online or watching television by three hours a day (from an admittedly gob-smacking 7 hours 38 minutes in the US).[124] Even if the message fails to sink in, technology can play a part, via wearable devices such as wristbands that monitor sleep patterns and send the data to the user's smartphone. Soon, children could be getting gold stars for their sleep rhythms as well as their homework.

And if all else fails, the mountain can come to Mohammed. Russell Foster worked with the headmaster of Monkseaton School, on the outskirts of Newcastle, to change the start of lessons from 8.50 a.m. to 10 a.m. The idea was that its sleep-deprived pupils (already biologically programmed to sleep in) could get some extra shut-eye. The results were impressive: the number of pupils meeting the government's required minimum thresholds at GCSE went from below the national average to above it. For children from disadvantaged backgrounds, the improvement was even more dramatic from approximately 20 per cent getting decent grades to just under 40 per cent.[125]

It is not just sleep. We now know that certain behaviours, or habits of mind, are as toxic as smoking cigarettes or drinking too much – and if we can avoid them, there is a huge dividend to be reaped in terms of health and happiness.

Take the world of work. Yes, it is plagued with distraction – but there are all sorts of tools we can use to control our worst instincts, such as carving out internet-free time or using time-management techniques to prioritise what matters and ignore what doesn't. Tools for self-discipline include setting up systems that will donate to a hated cause (the Republican party, say) if you fail to meet your stated goals, or which seek to counteract the way in which we discount the future to focus on the present – another key effect of the great acceleration.[126]

Even our smartphones can be conquered: the ethically sourced 'FairPhone' features a built-in option, prominently displayed, to turn yourself off from the world. 'We become more conscious of our phones as our lives become increasingly connected,' says the screen. 'Would you like to disconnect, if only for a moment?'

Indeed, it has been possible recently to determine the glimmer of a backlash against Schulte's Overwhelm. Blue-chip firms such as Atos and Volkswagen have either gone email-free (finding in-house social media networks more efficient) or turned off corporate email overnight;[127] Gothenburg in Sweden is experimenting with a six-hour workday in the public sector.[128] The giant American supermarket Best Buy embraced a programme called 'Results Only Work Environment' (ROWE), meaning that it didn't matter how long you were in the office, or where you did the work, as long as it got done. The outcome was a healthier, less stressed, more energetic and more loyal workforce.[129] Treehouse, an education start-up, positively insists on a four-day week, on the grounds that tired staff are useless staff.[130]

Too often, such practices are presented as a sign of weakness – an attempt to reject and deny the demands of globalisation and acceleration rather than to cope with them more effectively. When Best Buy, for example, changed chief executives, the new boss promptly canned the ROWE programme as too effete for the hard-charging environment he wanted to create.[131] Michael Arrington, the abrasive tech investor and founder of the TechCrunch website, summed up the prevailing culture in Silicon Valley in particular in a widely shared piece: 'Startups are hard. So work more, cry less, and quit all the whining.'[132]

Yet if you believe in the market – and in the evidence – then you are bound to believe that those firms that make the break, and ease their workers off the treadmill, will profit from doing so, to the point that the practice will inevitably spread, despite the inevitable resistance and disbelief.

This is not to advocate slacking off, but rather making the most of one's cognitive resources. Schulte calls it 'the power of the pulse' – working furiously when your system is primed to do so, then spending the downswing recovering.[133] Since the body's alertness oscillates in 90-minute cycles, this should grant ample time for productive effort. And in one of the slack periods, why not try going for a run? Exercise, after all, has a marvellous power to dissipate stress hormones (hence the title of one of Sapolsky's books, *Why Zebras Don't Get Ulcers*).[134]

Mastering the great acceleration is not just about how long we work – it is about the way we use our minds. As the great American psychologist William James said, 'the faculty of voluntarily bringing back a wandering attention, over and over again, is the very root of judgment, character, and will'.[135] Indeed, it turns out that executive attention – the ability to control our own impulses and focus our minds – is probably the single most important quality with which we should hope to be imbued.

We know this thanks to a remarkable study carried out in Dunedin, New Zealand, described by Daniel Goleman in *Focus*. The city – nestled on the southern tip of the South Island, and resembling (thanks to its Scottish colonisers) nothing so much as a boutique version of Edinburgh – has a population large enough to be statistically significant, but small enough to be trackable. So researchers tested every single child in a single academic year on a range of intelligence measures, then came back 20 years later to see what had happened to them. What they found was that the strongest single predictor of financial success – more than family, social class, or even IQ – was the ability to control one's impulses.[136]

That would have come as no surprise to Walter Mischel. Mischel is the American scientist who conceived the famous 'marshmallow test', which consists of leaving a child in a room with a marshmallow, and telling him or her that if they can refrain from eating it while left without adult supervision, they will be given two marshmallows as a reward.[137]

Children who pass the test, as with their equivalents in Dunedin, turn out to be happier, healthier and better adjusted than their peers – so much so that Mischel was brought in to help with a segment on *Sesame Street* in which Cookie Monster (and by extension the pre-school audience) was taught impulse control as a member of the 'Cookie Connoisseur Club'.[138]

The body of similar evidence is almost endless. Primates that receive training in attention, says Maggie Jackson, become less aggressive.[139] Six- and seven-year-olds who score high on effortful control are 'more empathetic, better able to feel guilt and shame, and less aggressive'.[140] And 'people who focus well report feeling less fear, frustration and sadness day to day, partly because they can literally deploy their attention away from negatives in life'.[141] Frank Partnoy, in his book *Wait*, points out that children who are fastest at regulating their heart rates after a shock also have the fewest social problems.[142]

The key point here is not that focus and self-control are good things, although they very obviously are. It is that they are neither finite nor innate: they can be trained and built up with surprising ease. There is no need to resort to mind-sharpening drugs such as modafinil: meditation classes and mindfulness lessons, for example, teach the brain to wander, then to return; apps and computer games can be devised that calm the players and help them regulate their instincts. Goleman describes how pupils in some special-needs classes are given 'breathing buddies' – cuddly toys which they clutch during calming sessions.[143]

In schools, the description for this is 'social and emotional learning' – and in Singapore, one of the most advanced economies on the planet, it is now compulsory.[144] In the KIPP charter schools in the US, students wear T-shirts emblazoned with the logo: 'Don't Eat the Marshmallow'.[145] Goleman points out that technology can also help reinforce our willpower, in the form of the 'biodot' – a small device monitoring your bloodstream that changes colour when your stress levels rise, allowing you the time to calm and collect yourself.[146]

The good news does not end there. Those who have high executive control are also able to develop another key characteristic – what might be called 'grit', the ability to stick to a task despite disappointment. This is, John Wayne would be unsurprised to learn, a better indicator of future success in life than SAT scores or IQ.[147] Crucially, developing these skills also makes us feel happier – which in turn delivers all of those things that we associate with happiness in the first place, such as better relationships, better jobs and better health.[148]

The final, and most convincing, proof of the benefits of such training comes from the world of neuroscience. Emily Ansell and her colleagues may have proved that stress shrinks the brain – but other researchers have proved that it grows back. Schulte describes how, when Bruce McEwen of Rockefeller University put rats into restraints for several hours a day for up to three weeks, he found that the prefrontal cortex and hippocampus, which govern learning and memory, had atrophied while the amygdala, which controls immediate emotional responses, grew.[149] But when he stopped the tests, the young rats recovered within three weeks (although for older specimens, the damage was more severe).

At Harvard, Britta Hölzel and her colleagues found that putting human lab subjects through eight weeks of mindfulness training led to increases in the density of grey matter within key areas of the brain – those involved in learning and memory, emotional regulation and perspective-taking.[150] In another study, she and others found that reducing stress – through a similar eight-week programme – was enough to produce positive structural changes in the amygdala, the seat of our panicked, fight-or-flight responses.[151] It was, they boasted, 'the first study to demonstrate neuroplastic changes associated with changes in a measure of a psychological state'. Similarly, adults who engage in five one-hour sessions on attention control show brain patterns closer to those of young adults.[152] There are even suggestions that such mental exercises – especially waiting for 15 seconds before performing any action – may be able to help OCD sufferers and habitual worriers by carving out new neural pathways that avoid tripping the 'alarm circuit' that sends them into obsessive behaviours.[153]

The lesson here is that yes, the brain is massively plastic – but that can be a force for good, too. If the great acceleration is damaging our brains and bodies, it is damage that can be fixed. Even Professor Wolf, who helped to foster the Slow Reading Movement, was able to retrain herself to read that Hermann Hesse novel within a remarkably short amount of time.

There are, therefore, reasons to be optimistic about our physical and mental capacity to cope with this great acceleration – and our ability to escape the societal self-destruction that many foretell. For example, living in cities floods our body with harmful cortisol – but it also makes us massively more productive, creative, prosperous and happy. One reason that the effects of acceleration are often so insidious is that they are so welcome – because they make our lives more fulfilling and exciting.

A basic function of the great acceleration is that it makes life speedier, and the demands on our time more intense. This panders to some of our most basic biological cravings – but in a way that is more usually invigorating rather than ruinous. We are all a little more stressed, a little more wired, a little more hyperactive. But as we shall see, we still have time for our families, for our passions, for long books and engrossing box sets. What we need most is not to slow things down, but to develop the right strategies to cope.

3

Fast Friends

'True love is a lack of desire to check one's smartphone in another's presence.'

– Alain de Botton[1]

Many people have argued that the acceleration of society is setting us on the road to degeneracy and ruin. But few have seized on the most obvious proof: the transformation of the TV dating show.

Once upon a time, at the birth of the genre, these were relatively sedate affairs. In *Blind Date*, or its US inspiration *The Dating Game*, a matronly/avuncular host would shepherd various youngsters through the perilous meadows of romance. The gimmick was that, being unable to see their would-be paramours, they were forced to judge them by words alone – even if those words usually consisted of horrifically laboured pick-up lines: 'Bachelor No. 1 – if you were a drink, what kind of drink would you be?' 'Well, I'd be a Guinness – because I'm strong, and dark, and if you give me two minutes, I'll give you a taste of heaven.'

At the time, such brief encounters seemed a rather flimsy basis for a relationship. Yet compared with their modern equivalents, they feel like an Athenian colloquy.

Since 2010, Saturday nights on the UK's main commercial network, ITV, have been graced by the appearance of *Take Me Out*. This is speed dating reduced to its most brutal: a series of young hunks descend into a bear pit of a studio, ringed by 30 girls who have been glossily styled to embody a single reductive characteristic: the brainy one, the posh one, the ginger one, the chav. As soon as the girls catch sight of the male contenders, they start hitting the buttons that switch off the lights in front of them, signifying their lack of interest. 'No likey, no lighty!' cackles the laddish host.

If all the lights go out, the man is unceremoniously booted off the show. But if more than one light remains by the end of his time on stage, the power dynamic flips: he is allowed to ask a single question of the girls who have indicated their interest, before slamming off the lights of those he rejects and heading off with his intended for the traditional caught-on-camera weekend of sun, sand and, hopefully, sex.

Take Me Out may seem like harmless fun, but in its unappetising combination of meat market and assembly line it speaks to some of our deepest concerns about how the great acceleration is changing our social lives. As our brains and bodies rewire themselves in accordance with the needs of a speedier culture, so the ways in which we interact with each other change. And in every sense, the governing principle of this new world is the quickening of desire.

The consequence, to the critics, is that kids grow up paying more attention to their iPads and televisions than to their parents, shunning the great outdoors in favour of flickering screens. By the time they have become teenagers, their focus has shifted. Inside the home, they lock themselves in their bedrooms, gossiping and flirting with their friends on one screen while playing fast-twitch, ultra-violent computer games on another. Outside it, they are never without a mobile phone, which serves as the conduit for all sorts of unsavoury activities – cyber-bullying, 'sexting', accessing a universe of inappropriate online material.

As they grow up, these fragile, narcissistic teenagers turn into fragile, narcissistic young adults, who are happier making emotional connections with machines than with people, and incapable of forming lasting commitments to their jobs, to friends, or to romantic partners. Instead, they live in a hedonistic blur in which the consequences of their drinking, their debts, their tattoos and piercings can be safely left to the dim and distant future.

Yet there is another description of the accelerated generation that is no less valid. These are children who are far more strait-laced than their wayward parents – they drink less, they smoke less, they have less sex, they knuckle down.[2] Going online also keeps them out of trouble: in 2007, 111,000 children under 18 were convicted of a crime by police in England and Wales, but by 2013, that had fallen to 28,000.[3] And these children are using technology not to abuse each other, or escape their daily life, but to enrich it, not least by maintaining and coordinating with a wide network of friends. They get harmless crushes on members of boybands, or cute YouTubers with adorable fringes; they create their own content and share it eagerly with their peers. They are less materialistic than their parents, more socially liberal, completely at ease with modernity.

Which portrait is the true one? Statistically, the two are equally valid – and most if not all children of the accelerated age are likely to have traits from both. Indeed, there is an 'Ever-Waser' case that, for all the doom-mongering, young people are basically the same as they always were. 'To sit there and argue that this generation is different from a previous generation is to ignore history,' Eric Schmidt, the executive chairman of Google, told me.

> Read the coverage in the Sixties about the King's Road, and the pop revolution. You want to talk about narcissism? Those people all grew up, and they're all sixty-five or seventy years old, and they all seem to have perfectly fine lives . . . Last time I checked, every young

> generation has had a narcissistic component, and somehow it works itself out, primarily through reproduction. When they ultimately have children, they actually have to care about someone else.[4]

Still, that can't be the whole story – otherwise we'd still all be watching *Blind Date*. Clearly, something about how we socialise must have changed over those years. But what?

To get a proper answer to that question involves wading through a whole lot of surveys and statistics. So I've chosen to borrow a device from the *New York Times* columnist David Brooks and his book *The Social Animal*, and create a fictional, prototypical child of the accelerated generation.[5] What sets her apart from her parents? What's so special about her life? First, we'll use her to look at modern childhood and youth – before turning our attention to the thornier issues of dating, sex and marriage.

THE LEGEND OF ZELDA

I've chosen to call my fictional girl Zelda – so named because, like so many other children these days, she's a bit of a princess. Her parents had her quite late on, and are doting and devoted. The problem is that they don't have as much time for her as they'd like, because they're already stretched, and stressed, to their limits (maybe it would be different if they'd taken those mindfulness lessons . . .).

As a result, Zelda's are among the three-quarters of parents who feel they don't spend enough time with their children, and the two-thirds who don't have enough time for themselves or their partners.[6] It's not that bad – Zelda's father would never dream of cheating, but mostly because he just can't see how he'd find the time (in the UK, adultery is now cited in only 14 per cent of divorces, whereas workplace pressure is a factor in more than half).[7]

Between his job and his wife's, and the demands of looking after Zelda and her brother, everyone in the family is feeling the strain.

Family meals together, or even quality time in the same room, are getting rarer and rarer: the average family now spends just 49 minutes together a day (this, indeed, is one cause of kids' problems with etiquette and manners: they spend too little time at the dinner table to be socialised).[8] It's got to the stage where Zelda and her mother don't even look up to say hi when Zelda's dad comes home from work.[9] When she was younger, her dad bought Zelda a bear that offers long-distance 'tele-hugging', so he could let her know how much he loved her while he was still in the office. She thought it was the creepiest thing she'd ever seen.

So what fills the parenting void? In a word, screens. Zelda and her generation are the first real children of the digital age, surrounded by information almost from the day they were born. By the time she hits seven, Zelda has already racked up a full year of screen time – if you double-count time spent using more than one device.[10] Her parents can't believe they actually thought it was cute when she started swiping at the glass frame of their wedding photograph, or the bathroom mirror, in imitation of Dad and his iPad – and still regret how quickly they gave her her own little tablet, far earlier than they'd ever expected to. In fact, they're starting to worry. Surely it can't be normal, they fret to the former members of their NCT class, that their five-year-old daughter can happily play all sorts of computer games, but can't tie her own shoelaces?[11]

In fact, there are two factors which will determine – or at least shape – the kind of child Zelda grows up into. The first is whether staring at screens is all she does, or whether she gets out into the real world for playdates as well. These are vital for various reasons. For one thing, physical activity is the best possible way to shake off the mental drag of screen time, just as it is for adults. For another, some scientists have linked the growth of narcissism mentioned in the previous chapter, and the decline in empathy (which has fallen among US college students by 40 per cent in recent years), not to the dominance of online communication, but to the absence of the offline kind.[12] For it is in interacting with each other, face to face, that we stimulate the mirror neurons that teach us to mentally mimic, and appreciate, others' thoughts and feelings.

Another key question is not whether Zelda is looking at screens, but which kind of screen she's looking at: to use the shorthand, is she leaning forward or leaning back?

In Zelda's case, her mother learned early on that switching on *CBeebies* or *Nickelodeon* was the easiest way to get a moment's peace. Unfortunately, this may be the worst thing she could have done. With the possible exception of the internet, TV is the greatest attention-snatcher ever devised. And, as Brigid Schulte writes, studies have shown that TV is 'making us fat, depressed, socially isolated, and more prone to violence; lowering our self-esteem, disrupting our sleep, dulling our senses, fogging our mind, and shortening our attention and life spans'.[13] Within only 30 seconds of it flicking on, we effectively lose our sense of self – our thinking brains switch off, and our alpha waves 'become no more active than if we were staring at a blank wall'.[14]

Television is perhaps the perfect incubator for children of the accelerated age: it wires up their plastic young brains to expect and crave constant, rapid changes, to the point where reality starts to seem boring by comparison. Why, for example, are Ritalin prescriptions in Britain up 300 per cent in a decade?[15] Well, children aged one to three will play with toys for half the amount of time if a game show is on in the background, and show up to 25 per cent less focus in their play.[16] As Maggie Jackson writes in her book *Distracted*, such kids 'exhibit key characteristics . . . of attention-deficit children. They begin to look like junior multitaskers, moving from toy to toy, forgetting what they were doing when they were interrupted by an interesting snippet of the show.'[17]

Experts have calculated that every hour of TV watched per day before the age of four bumps up your chance of getting ADHD.[18] This, indeed, is the quintessential modern disease: the psychologist David Gilden of the University of Texas has claimed that its root cause is not an absence of attention, but a distorted sense of time – that sufferers are literally moving too fast.[19]

The prevalence of TV also means that parents interact less with their children, to the point that children in houses where it is always on generally prefer watching it to spending time with their mum or dad.[20] Furthermore, increased TV viewing is, according to Public Health England:

> associated with lower self-worth and self-esteem and lower levels of self-reported happiness . . . children who spend more time on computers, watching TV and playing video games tend to experience higher levels of emotional distress, anxiety and depression. This relationship is particularly negative among those who engage in high levels of screen use (more than four hours a day).[21]

To put that in perspective, the Kaiser Family Foundation has found that Americans aged between eight and eighteen use various media for 7 hours and 38 minutes per day, or more than 10 hours if you count overlapping screens – far more than the danger limit.[22] And even if you are limiting your children's exposure to sensible levels (something more easily said than done) there is still the issue of policing the type of programming. In one study by Angeline Lillard and Jennifer Peterson at the University of Virginia, four-year-olds who watched nine minutes of *SpongeBob SquarePants* (admittedly, a show aimed at older children) suffered an immediate negative impact on their executive function – they found the fast pace and fantastical events confusing, and didn't have enough time to assimilate the new information.[23]

GROWING UP FAST

Zelda doesn't have ADHD – but she's certainly watched more TV than is strictly good for her. She's also growing up faster than her brother – by the age of eight, she's already pronounced *Moshi*

Monsters 'lame' and moved on to more sophisticated pursuits, whereas he was still using the site right up until the age of 12. But she's still basically a good kid.

And in fact, the presence of screens in her life is starting to help rather than hurt. As she grows up into a teenager, Zelda will start to use the internet obsessively, largely to chat and catch up with her friends. But the time she takes to do this won't come out of her time with her friends, or her family – it'll come out of her TV viewing, to the point where she watches far less than her parents.[24] Zelda's father has even taken to using the TV as a punishment: rather than turning it off, he forces her to watch it instead of using her mobile or tablet when she's been naughty.[25]

Increasingly, the only way Zelda finds the TV tolerable is when she's also got her mobile in her hand – largely to send catty comments about what she's watching. (A study in 2011 showed that 97 per cent of 16- to 24-year-olds 'always' or 'frequently' go online while watching television.)[26] Again, her parents fret about her lack of attention and inability to focus, and there's something to that. But she's also more diverted, stimulated and engaged than she would be just from watching the TV: brain studies (admittedly carried out by Twitter itself) have shown huge lifts in neural activity from such multi-screening.[27]

Zelda is, of course, an awful teenager – but then again, weren't we all? The sloshing hormones of puberty (which now comes earlier than it used to, compressing childhood further) and the lack of sleep we discussed earlier are certainly wreaking havoc. Her parents' failure to set simple rules about media consumption – for example, that mobile phones must be placed in the middle of the table during meals – have helped consign her to digital distraction.

At least, that's what her parents assume when they're ever allowed to enter her bedroom, which seems to have almost as many screens and chargers as it does posters of boy-band heart-throbs. With them, she communicates in the normal surly teenage grunts – but with her friends, she fires back and forth a blistering array of instant messages, pictures, jokes, opinions. The very idea of conversation, her parents worry, seems to have been replaced by an electronic blizzard of half-formed half-thoughts. It surely can't be healthy, they think, that Zelda and her friends text each other more in a day than her elders do in a year.[28]

Part of this mutual incomprehension cannot be helped. One consequence of the technological churn that powers much of the great acceleration – the constant hunt for new and disruptive business models – is that communications tools are changing more rapidly than ever. When Zelda and her friends are using services like Snapchat, which didn't exist even when their older siblings were being raised, what hope have their parents got of bridging the gap?[29]

Much – indeed, almost all – of what Zelda is doing online is actually entirely harmless (well, apart from joining in with the occasional bit of cyber-bullying, but honestly, did you SEE what Julie was WEARING?). But she still has a problem. Several problems, in fact.

First, she is being trained by that constant stream of messages – hundreds a day – to expect and crave information and distraction, and its inbuilt dopamine hit: the Skinner box in action. The next problem, if Zelda isn't careful, is that she'll start to view technology not as a means to communicate with other humans, but a substitute for it.

One of the most powerful critiques of the effect this has on our emotional lives comes from Sherry Turkle, a leading US sociologist, in her book *Alone Together*.[30] 'These days,' she writes, 'insecure in our relationships and anxious about intimacy, we look to technology for ways to be in relationships and protect ourselves from them at the same time.'[31] This takes two forms. First, a desire to mediate our encounters with other humans through our devices (for example by sending instant messages rather than placing a phone call or meeting face-to-face – another reason, along with the greater speed and convenience of messaging, why voicemail and even phone calls are basically dying out).[32] Second, a preference for such machines' unthreatening company over that of human beings.

In Zelda's case, she may develop a thriving social life – or a few bruising dating experiences may see her retreat further into her bedroom. Certainly, we should hope she doesn't end up like some of the people Turkle has profiled. There was the woman who turned up to interview a new nanny, only for the interviewee's flatmate to open the door. Despite bandaged thumbs, she was still tapping away at a BlackBerry, and insisted on texting her friend rather than knock on her bedroom door: such a real-world intrusion would be dreadfully impolite.[33]

Luckily for Zelda, she has a good network of friends around her. Indeed, she may well be closer to them than her parents ever were to theirs: studies have shown that young people tend mostly to hang out online with the people they hang out with face-to-face.[34] The effect of the constant online interplay is to heighten feelings of closeness. Indeed, her parents often worry that Zelda's friendships – and especially her relationships – burn too brightly.

The problem is that, in an accelerated culture, digital children become friends faster than ever. As Dan Slater writes in *Love in the Time of Algorithms*, 'You begin dating someone you met online, or off' – after devouring any information you can find about them on the internet, of course – 'and in a matter of days you are Facebook friends who also follow each other's Twitter feed and show up on each other's Tumblr dash and chat throughout the day via IM and text. By midday you've opened ten tabs on your browser, and on five of them the avatar of your paramour is blinking and winking and typing and poking and accepting and liking and smiling and frowning and inviting.'[35] Not for nothing is Facebook known in Trinidad as 'Fasbook', meaning 'to try to get to know someone too quickly'.[36]

In Zelda's case, she's a bit of an egotist – how can one not be, when one uses social media? But she's also rather insecure: because whenever she posts selfies or other pictures online, she can't help but notice that everyone else looks so much better than she does. She's also a bit shy about romance – because there was this one girl in her class who sent a boy a saucy picture, like he asked, and he shared it with absolutely everyone. No thank you, thinks Zelda.[37]

Similarly, when she did have a boyfriend, she's afraid she messed it up, spending so much time following him round various social networks and worrying about who he was talking to who wasn't her, that it was hardly any fun at all. And when they broke up, there he was, all over her feeds, a constant reminder of what she got wrong.

Then there was this other boy, whom she met online. He seemed really nice when they started messaging each other but turned out to be a creep when they met in person. The problem there is that when we can't see someone, we imagine that they are similar to us, and share our values and emotions. 'Because we invest so much emotion, expectation and idealism in the objects of our virtual affections,' explains Aleks Krotoski in her book *Untangling the Web*, 'web-based love affairs tend to feel hyperintense, hyperquickly. And because there are so many blanks we have to fill in, we can over-enhance a partner's positive features and make up the rest to meet our romantic ideals.'[38]

Beyond her romantic misadventures, Zelda also has a deep, dark secret: that she's actually not much of a party animal at all. Her Facebook feed – not that she spends much time on it, but if she didn't have one, her parents would wonder where she was actually spending her time online – is carefully packaged to present a cheerful, upbeat, outgoing image: lots of pictures of her at parties, or out with her friends.

Yet those same apps that tell her where her friends are all the time, and let her broadcast her thoughts to the world, also let her know what she's missing out on. She can see where her friends are, and can notice from their Foursquare or Dodgeball check-ins that Tasha and Olivia and Jenny have been in the same coffee place – without her. If she tries to disappear from online media for a few days, to catch her breath, people ask her why the radio silence. Sure, her parents' generation knew there was always a cooler party somewhere – but they couldn't see the guest list.

So 'lifestreaming' – the practice of broadcasting your every move to your friends online – is often accompanied by crippling 'FOMO' – 'fear of missing out'. All these apps and notifications seem to always be urging you on to be hypermobile, hypersocial, to enjoy every moment without thinking about the long-term cost. Indeed, the addictiveness of checking your phone – fear of missing out on whatever messages may have arrived – plays a large part in undermining the enjoyability of social occasions for those of any age: how many meals or conversations have you had that have been disrupted by everyone surreptitiously, or not so surreptitiously, scanning for new messages?

But here's the paradox. From the outside, Zelda seems to be outgoing, connected, even slightly rootless. But actually, her social horizons are surprisingly narrow. The flipside to her slightly cautious character is a rather conformist nature: as you may have picked up, she's not much of a risk-taker. Even after leaving university, she's stayed with her parents – partly because she can't afford a place of her own, but also because, however distant they were while growing up, they're still the people she's closest to. In fact, it was only the regular phone calls with her mum (twice daily, almost) that got her through university.[39]

This may seem strange to some. While most online interaction is, as we have seen, a continuation of and a buttress to offline relationships, it also gives us the opportunity to expand our social circle, to find those who share our interests or passions with far greater speed than ever before. As Krotoski says, that is good for everyone: 'Almost every research study . . . has pointed in the same direction:

online friendships are not only possible, but people feel less socially isolated when they go online ... the web reinforces friendships with people we already know and people we've just met.'[40] This process also widens our horizons. Eric Schmidt recalls: 'Until I graduated from high school, I had only met one Jewish person – it was that homogenous, the world I grew up in in Virginia. Like, the one Jewish person was quite notable. Think about how much narrower my world was.'[41]

Yet for many people, the uncertainties of the great acceleration also provoke the opposite reaction – to retreat from the world into the comforting, safe and familiar. And that's the case even more with Zelda's poor male friends – after all, a social scene dominated by networking and social media is one in which unmasculine values like sharing and sympathy and emotional literacy are privileged, and the poor, confused, hormonal men are all too often dismissed as sex-obsessed and emotionally illiterate, when actually they're just shy.[42]

By now you may be slightly confused. Wasn't Zelda supposed to show how and why today's youth culture went so badly wrong? Instead, we've ended up with a girl – well, a rhetorical device based on an averaging-out of modern family experiences – who's as much an introvert as an extrovert. Certainly, she couldn't be further from lurid stories about a 'hookup culture', in which young people treat sex and intimacy as an à la carte menu rather than a package deal.

The explanation, of course, is that there is no one set reaction to the opportunities that acceleration offers. Some embrace their wildest possibilities; others – like our Zelda – seem to be at home with them, but nurse secret doubts and fears; still others grow up much as their parents did.

But before we leave Zelda be, there's one more question to ask. Are her problems, such as they are, the fault of her accelerated life-style? Or do their root causes lie rather closer to home?

The truth is that the more you look at her life, and the effects of the great acceleration on it, the more you realise the truth of Philip

Larkin's dictum: 'They fuck you up, your mum and dad. They may not mean to, but they do.'[43] In particular, they do so not by trying too little to make their daughter happy, but too hard.

We talked in the last chapter, for example, about the need for children to develop grit. One of the prerequisites for that is that they are left alone to make their own mistakes: to experiment, to fail, to struggle, to find the things they love to do. But in a world of helicopter parenting, where each child is seen as a precious jewel, this is the last things parents want to do – even parents as stressed and distracted as Zelda's.

For today's children, as we saw in the last chapter, their spare time is over-structured, packed with extra lessons and activities.[44] That's not just because parents are desperate to spur their development – it's also because they're frightened of letting them fly free. In her first-hand study of modern teendom, *It's Complicated*, the US sociologist (and capital-letter-phobe) danah boyd points out that in America, half of kids walked or cycled to school in 1969, and only 12 per cent drove; by 2009, those proportions were almost exactly reversed.[45] In Britain, the proportion of seven- or eight-year-olds walking to school dropped from 80 per cent in 1971 to just 9 per cent in 1990.[46]

And what about all that time teenagers spend shut in their bedrooms, firing messages back and forth? According to Howard Gardner and Katie Davis, authors of *The App Generation*, much of that back-and-forth consists of 'making (and sometimes breaking) on-the-fly arrangements to meet up with their friends in person'.[47] It's the friendships, not the devices, that matter to them – the latter are, by and large, a means to an end.

And why aren't they out there meeting in person? Because, boyd points out, the adult world seems to have dedicated itself to removing any unsupervised spaces in which teenagers can hang out: even when they loiter in shopping centres, they tend to be evicted by security guards. 'Facebook, Twitter and MySpace are not only new public spaces,' she says. 'They are in many cases the only "public" spaces in which teens can easily congregate with large groups of their peers.'[48] Similarly, the behaviours about which adults fret – sexting, cyber-bullying and so on – are by and large outgrowths of offline behaviour.

It turns out that those children who have severe problems, who are driven to internet addiction or suicidal thoughts or inappropriate relationships with anonymous online strangers, are mostly those whose home lives are genuinely troubled.[49] It is another example of what we saw in the last chapter – the way the great acceleration preys on the weakest. And the same phenomenon is apparent in adults: over the last 20 years, the proportion of Americans with no close confidants at all has doubled to a quarter.[50] The worry is that, even though we have plenty of weak social ties, we are left – if we are already vulnerable – with few or even no strong ones. But all this does not mean that the fast-paced social lives of these young people are dangerous or ill-advised in and of themselves – it just means that we need to make greater efforts to be aware of and support those who are most vulnerable.

THE CURSE OF OVERSHARING

One of the most common complaints about our culture is that in the faster world of the internet, niceties like manners and politeness are dead weight. Even the language we use to talk to each other is being transmuted, stripped of meaning and complexity.

'In the electronic age, everything has to be accelerated,' laments Henry Hitchings in *The Language Wars*. 'Words, which have to be used in a linear fashion, frequently cede ground to images. On the internet capital letters and verbs are apt to disappear, along with punctuation – question and exclamation marks are retained and indeed sometimes laid on extra-thick, but commas evaporate . . . communication is neither speech nor writing as we conventionally understand them, and it is staccato.'[51] Others, such as John Humphrys, lament that new forms of communication are 'doing to our language what Genghis Khan did to his neighbours 800 years ago'.[52] And that was even before emoji – those little picture symbols in text messages – became a shorthand all of their own.

Yet for all that we fret about textspeak or emoji, there is virtually no evidence that the internet is destroying our writing abilities – instead, millions more people than ever before are writing for a public audience, or just writing full stop, and in proper English too. Exhaustive analysis of millions of words written online, including crunching through every single word used on Twitter, has shown that our writing style is, if anything, getting better rather than worse, polished both by constant use and by the need for concise expression.[53]

Clive Thompson, in his book *Smarter than You Think*, cites the work of Andrea Lunsford, a Stanford English professor who analysed students' first-year essays over more than a century. She found that the number of typos had only risen slightly – but the essays' length and the complexity of thought had risen hugely. 'I think we are in the midst of a literacy revolution the likes of which we have not seen since Greek civilisation,' she claims.[54] It may be that, as the online communications revolution rolls on, a shift from text to video undoes that good work. But there is no sign of it yet – quite the reverse, in fact.

The critics' next line of attack is to move beyond the specifics of language, to argue that there has been an erosion of manners and politeness – both relics of the slow-moving world. We no longer get dressed up for dinner, or even for job interviews, just as we no longer bother to address each other as 'Mr' before moving gradually on to first-name terms – everything is fast, chatty, informal. 'The eloquence of letters has turned into the unnuanced sparseness of texts,' complained Rebecca Solnit in the *London Review of Books*. 'The intimacy of phone conversation has turned into the missed signals of mobile phone chat.'[55]

It is true that online culture is less hierarchical, and less deferential – though opinions may divide over whether that is such a bad thing. A more serious concern, however, is not the manner in which we communicate, but what we are actually saying. For example, damaging or hurtful material can spread far more quickly – think cyber-bullying, or pro-anorexia messages. Indeed, an entire 'pro-ana' culture has grown up, with Tumblrs obsessed with pictures of 'thigh gap' and forums in which people are told that skeletal thinness is perfectly normal and desirable. Pro-suicide sites are also a curse: just ask Martyn Piper, who became a campaigner on this issue after his son Tim killed himself at 16 by following the instructions he found online.[56]

That, indeed, is the downside of an accelerated culture: that bad ideas and influences can spread more quickly. And the opportunity for harm is all the greater because of the way in which, in this less formal age, we are willing to share more details about ourselves. For example, of those who answered Yes to the question 'Have you ever had an abortion?' on OKCupid, less than half ticked the box to keep their answer private.[57] Our willingness to expose ourselves is, in fact, growing more and more literal: according to pollsters YouGov, a fifth of British adults under 40 have engaged in sexual activity in front of a camera, and 15 per cent have appeared naked in front of a webcam.[58]

The trouble is that, when things go wrong in this new world, they go wrong in an instant. Recently a host of Hollywood celebrities found their personal pictures – including nude shots – spread across the internet after Apple's iCloud storage database was hacked, an episode known as 'the Fappening'.[59] And in his fascinating book *The Dark Net*, Jamie Bartlett of the think tank Demos describes a 'life ruin' on /b/, the most extreme and outrageous of the message boards on 4chan (and one of the original spawning grounds for the Anonymous movement).[60]

The story starts with a would-be 'camgirl', a young girl who decides to stage a peep show via webcam from the privacy of her bedroom. This is a growing phenomenon, to the point where it has disrupted much of the old-fashioned porn industry.

But this particular camgirl has made a mistake: she has popped up on /b/ rather than on /soc/, a safer and more respectful message board. She starts taking requests for poses or activities. Twenty minutes after posting her first photo, she makes the further mistake of revealing her first name, Sarah – and then of answering a request to pose with any medication she is taking.

This limited personal information enables her to be 'doxxed' – to have her anonymity stripped from her. On the board, a feeding frenzy develops. Members find her name, address, university, telephone number, Facebook and Twitter accounts. They create a fake Facebook account with a collage of her nude pictures, and send it to everyone on her friends list: 'Hey, do you know Sarah? The poor little sweetie has done some really bad things. So you know, here are the pictures she's posted on the internet for everyone to see.'

Just as shocking as the speed with which this happens is the basic lack of empathy. One anonymous user comments: 'You fucking nerdbutts got her Facebook? You guys are fucking unbelievable. A girl actually delivers on this shit site, and you fuckers dox her. Fucking /b/, man.' The response? 'Get the fuck out you piece of shit moralfag trash.' Someone gets her phone number, and announces: 'Just called her, she is crying. She sounded like a sad sad sobbing whale.' Within an hour, the thread has vanished, and Sarah is forgotten by /b/ – but her life lies in shreds.

Partly, this is a result of the 'disintermediation' effect – the idea that, because we cannot see the people we are communicating with online, we treat them with a brutality we would never dream of using face to face. 'Because you can't see the hurt in the other person's eyes, or them clenching their fists, online discussions simply tend to disintegrate,' laments Ben Hammersley, an editor at *Wired*.[61] This is a phenomenon best described by the webcomic

Penny Arcade's Greater Internet Fuckwad Theory: 'Normal Person + Anonymity + Audience = Total Fuckwad.'[62]

The consolation here is that the same anonymity that makes people more willing to hurl online abuse may also make them more honest, whether it is opening up to a non-judgemental computer program about their medical or emotional problems, or informing the sympathetic strangers in an online forum that they think they might possibly be gay. But as Zelda and her friends learned, in a society where everyone decides to let it all hang out, we're not going to like everything we see.

THE QUICKENING OF DESIRE

It is one of the most intriguing questions about the great acceleration. What happens when this Google generation carries out the most important search of all – for someone to share their life with?

The growth of the internet has been accompanied by an explosion in dating websites. Match.com sells itself as the world's matchmaker, with a near-inexhaustible stock of potential partners; eHarmony claims to use its extensive psychological profiling to find the one person whose psyche will fit, jigsaw-close, into yours; OKCupid is the data specialist whose algorithms will chew through thousands of fun quizzes to reveal, for example, that a shared taste in horror movies is mathematically the best sign of compatibility.[63] Beyond that lies a universe of more specialist sites: uniform dating, Jewish dating, nappy fetishist dating.

Not only is this a vast business, but it is now a thoroughly normal one – among upscale young professionals in London, for example, it is now distinctly abnormal to encounter someone, even a lifelong Tory, who has never used Guardian Soulmates. Between 2004 and 2014, a third of US marriages were between couples who met online.[64]

The growth of online dating fits neatly into the pattern of the great acceleration: it is obviously quicker and more convenient to scan through hundreds of online entries, or to have the best matches selected by an all-knowing computer, than to go through the laborious process of actually meeting people. This is exactly the reason why speed dating has gone from a curiosity to a mainstay of the dating scene: it is not only more efficient but also more variously enjoyable, offering a constant stream of novelty, and removing the risk that you will end up grinding your way through an entire evening with a date with whom you don't quite click.

And, as you would expect, the popularity of such forms of dating has indeed had an effect. As far back as 2006, the researcher Monica Whitty found that, year on year, we're becoming fussier about who we date – either paralysed by the choice available, or believing that if Mr Right really is out there, Mr Almost Right is no longer good enough.[65] The key to happiness, as in other areas where we are confronted by infinite choice and information overload, is to become 'satisficers', happy to make choices that are just good enough – but online, you can't help seeing what other options are out there. (In Italy, according to its divorce lawyers' association, infidelity via WhatsApp is now cited in 40 per cent of adultery cases.)[66]

We're also becoming more superficial – or perhaps having our innate superficiality revealed. While the technology writer Julian Dibbell has noted that online culture privileges wit and literacy – 'it's the best writers who get laid'[67] – that's only true if you can get people to look at your profile in the first place. And unless you're attractive enough, they won't. Christian Rudder, co-founder of OKCupid and its chief data scientist, describes how, when the site made the pictures larger, it jacked up the attention paid to the most beautiful people – another example of the winner-takes-all economy.[68] (He also showed that America is still tainted by racial prejudice: on average, being black costs you three-quarters of a star out of five on your attractiveness rating.)[69]

In fact, OKCupid has provided definitive proof of our insistence on judging by appearances. In one experiment, the site declared a 'Love Is Blind Day' and removed the profile pictures from its users' pages. Rather than exploring the written profiles in depth, as the site hoped, people simply stopped using it. It was their loss: it turns out that, for those who did go on the blind dates that resulted, their actual enjoyment was utterly uncorrelated to their partner's physical appearance.[70]

Is online dating making us more superficial in our love lives, then? Well, yes, as one executive at the rival dating site Badoo told Dan Slater: 'There's not a lot of text, not a lot of substance. It's about appearance. And everything's very quick. "Show me your photo, tell me how old you are, then let's connect." Badoo just makes the world work faster.'[71]

But actually, the data suggest that online dating isn't changing who we date, just how quickly we meet them. A 2010 study co-authored by the behavioural economist Dan Ariely found that dating patterns online 'are qualitatively similar to those observed offline'.[72]

In fact, the most interesting thing about dating sites – certainly from the point of view of the great acceleration – is not how successful they've been, but how ineffective. Yes, they match millions of people – but the techniques they use to do so, no matter how sophisticated, are still little better than guesswork.

The truth is that, however good your psychological profiling, the only thing that determines whether two people will get on in the long term is putting them together to find out. As Eli Finkel of Northwestern University says, 'Eighty years of relationship science has reliably shown you can't predict whether a relationship succeeds based on information about people who are unaware of each other' – no matter how much information you have about their taste in movies, or who they're friends with on Facebook.[73]

There are two personal stories which illustrate this neatly. Amy Webb and Chris McKinlay were both lonely – and, to be honest, rather strange – people who became frustrated at their lack of success on dating websites. So both decided to game the system.

Webb, a journalist, did this by transforming herself. In her book *Data, A Love Story*, she relates how she worked out what qualities the most popular girls on these sites had, and systematically remade herself and her profile to match (which, in something of a body-blow to feminist principle, involved cutting out all the stuff about her satisfying, fulfilling career and behaving like a bubbly, flirty blonde).[74] At the same time, she ruthlessly calculated which attributes she sought in a mate, and refused to date any men who did not approach her idea of perfection.

McKinlay, a maths nerd, took a more elemental approach: by setting up 12 fake accounts on OKCupid manned by bots, he was able – through brute-force interpretation of the data – to build up a profile of the female users.[75] These divided, he found, into seven clusters, two of which appealed to him – creative and artistic women in their mid-twenties, and slightly older women working largely in editorial and design. So he set up two personal pages custom-designed to appeal to those groups, stressing different aspects of his personality (his passion for rock-climbing, his guitar-playing, his teaching career) according to that particular market's interest.

Similarly, while McKinlay was scrupulously honest in answering the endless quizzes that OKCupid peppers its users with (the better to collect data on them), he weighted the importance of his answers. When asked whether he thought love mattered more than sex, he said yes both times – but for the cluster of younger, cooler girls, he said that was 'very important' to him in a relationship; for the older group, he said it was 'mandatory'.

Webb and McKinlay, in their separate ways, gamed the dating websites to their advantage: both saw their position on the site rocket upwards (McKinlay, on launching his new profiles, moved from having fewer than 100 women ranked with a compatibility rating above 90 per cent to more than 10,000). Yet both of them still found it an exhaustingly long process, even when faced with the very best fish in the sea, to find their perfect match – 88 separate dates, in McKinlay's case. What mattered, as ever, was being in the same room.

And this is the vital flaw in traditional online dating. It promises greater speed and convenience, but it doesn't really deliver. Yes, it saves you the time spent on wasted first dates, where you know even before ordering the starters that it's not going to work out. But it's still not as efficient a way of meeting The One as it could be. In a study Dan Ariely conducted, he found that online daters spent an average of 5.2 hours per week trawling through profiles, 6.7 hours emailing potential partners – and just 1.8 hours actually meeting people, which in most cases consisted of 'nothing more than a single, semi-frustrating meeting for coffee'.[76]

To make matters worse, the whole experience was rated by its users as being about as fun as pulling teeth. 'Online daters aren't particularly excited about the activity,' says Ariely. 'They find the search process difficult, time-consuming, unintuitive, and only slightly informative. Finally, they have little, if any, fun "dating" online. In the end, they expend an awful lot of effort working with a tool that has a questionable ability to accomplish its fundamental purpose.'[77]

THE SWIPE IS RIGHT

Online dating, in short, was ripe to be disrupted. And it has been, by services that are even more convenient. The first is the large social networks, not least Facebook – the idea being that online dating is so normal that it is becoming simply an offshoot of regular online interaction, with no need for special external venues.

But there is an even more popular attempt to make online dating a seamless part of life: the hook-up app. The most popular is Tinder, which is essentially an updated version of the gay sex app Grindr.

Rather than finding you the perfect match, this service uses your mobile phone's location technology to flash up an endless stream of nearby individuals, who can be swiped through as easily as shuffling a deck of cards. If they like you too, then – alleluia! – your phone buzzes to tell you you've been matched.

This app is the perfect encapsulation of so many aspects of our accelerated culture: instant sexual convenience, delivered straight to your mobile phone. And it has spread like wildfire. Yet for its detractors, Tinder is a classic example of why an accelerated culture is also a degraded one: it strips romance from the equation, replacing it with a sequence of semi-random hook-ups, exactly like a computer game. This date didn't work out? Just press continue: or, as the app actually asks, 'Keep playing?' Indeed, the near-hypnotic ease of flicking through the options is a key part of the appeal: some of Tinder's rivals even adopt the tricks honed by free-to-play iPad games such as *Candy Crush* to maintain your interest, such as countdown clocks until a new batch of matches are released.

This is the serious business of life and romance, reduced to base entertainment: *Take Me Out*, played every night by an entire generation. Rather than true romance, it delivers instead a stream of instant, fragmentary experiences. No swipey, no likey. And, as ever, its impact has been portrayed as apocalyptic. In a lengthy survey of the Tinder-driven dating scene in *Vanity Fair*, Nancy Jo Sales depicts a landscape of 'fuckboys' and hook-ups, in which men dial up a sexual partner as easily as an Uber taxi, and commitment and caring have been replaced by an endless cascade of sexual novelty, with messages flying back and forth along the lines of 'I want to have you on all fours' or simply 'Wanna fuck?'. Sending dick pics is commonplace ('It's like we have dicks flying at us,' complains one 24-year-old girl), as are hook-ups conducted entirely in emoji. 'When it's so easy, when it's so available to you,' says one of Sales's interviewees, 'and you can meet somebody and fuck them in 20 minutes, it's very hard to contain yourself.'[78]

Yet there's another way of looking at Tinder. Not only does it make it easier to find prospective partners – whether for long-term or short-term relationships – but it's actually a great example of people using digital tools and technology to enhance their experience of real life: to make it more rewarding and enjoyable, rather than retreating into their phones and shunning real-world interaction. Certainly, it's hard to imagine previous generations turning down the chance to use such tools if they'd been available.

As ever, what technology is changing is not human nature, but the opportunities to express that nature. And even on these terms, the idea that we are forsaking long-term relationships in favour of a series of mutually satisfying yet ultimately shallow encounters really doesn't hold water.

Certainly, there are signs that the two great gains of acceleration – novelty and convenience – are affecting what we want from our love lives. The writers Jessica Massa and Rebecca Wiegand have described how young women increasingly maintain a 'gaggle' of male friends, each with his own role (bedmate, confidant, household repairman), rather than choose a single paramour.[79]

There is also some debate as to whether the stream of novel partners and experiences that dating websites offer will provide a greater attraction, over the long term, than a single settled partner – in other words, whether the balance between promiscuity and pair-bonding will tip in favour of the former. Sales quotes a researcher called Justin Garcia, who claims the shift in dating behaviour driven by the internet will be as significant, as epochal, as the shift towards marriage 10,000 to 15,000 years ago.[80] Carl Djerassi, creator of the Pill, predicted before his death that we will soon separate love and sex, spending our twenties frolicking, then having children later on, once we are ready to settle down (by means of sperm and eggs frozen when we were young).[81]

There is a marvellous experiment involving rats that suggests there might be something to such ideas. In this instance, when scientists put male rodents in cages with females, they have sex until the male is apparently exhausted. But put a second rat in, and they suddenly recover their libido. So strong is the lure of novelty, in fact, that they can repeat the performance when a third female appears.[82]

For the executives behind some dating websites, our rodent-like craving for new partners will indeed overpower our hidebound morality. 'If dating through the Internet becomes more and more popular, and sites become more efficient, what do you think will happen to commitment when people discover how much easier it's become to find new relationships?' Dan Slater asked Sam Yagan, co-founder of OKCupid. His answer? 'That's a point about market liquidity, which I really like.'[83] 'It's exhilarating to connect with new people,' agrees Nic Formai of Badoo. 'Over time, you'll expect that constant flow.'[84]

Noel Biderman is the founder of Ashley Madison, which sells itself as the online home for extramarital liaisons, using slogans such as 'Monogamy is monotony' and 'Life is short. Have an affair' (and suffered the indignity of having its entire client list stolen by hackers – another example, in the accelerated age, of how quickly things can turn sour).[85] For Biderman, when technology collides with cultural norms, those societal values 'always lose out'. As a result, he told Slater, 'the old thinking about commitment will be challenged very harshly' over the next few decades.[86]

The problem is that the evidence contradicts him – not least the fact that only an astronomically small proportion of the regular users on Ashley Madison turned out to be actual human women.[87] For all the dizzying temptations of Tinder and the like, even those who rejoice in the disposability of these attachments still hope and intend to settle down eventually.[88] Yes, the great acceleration has delayed family formation – but that is in part because it has produced high-pressure workplaces that push women to delay bearing children.[89] Similarly, the reason that fewer and fewer young people have moved away from home and become independent isn't that they're so diverted by the instant pleasures of the hook-up culture – it's because they simply can't afford it.

In their book *Jilted Generation*, my friends Ed Howker and Shiv Malik showed that much of the delay in family formation, at least in the UK, can be explained by sky-high house prices – since we are all understandably reluctant to settle down and have families until we can afford a place of our own.[90] And the fact that people are spending their twenties frittering away their cash on wild nights out – often adduced as evidence of a short-termist, ADHD culture – is at least in part a rational response to the fact that home ownership, in cities such as New York or London, is increasingly an impossible dream, so saving for it is essentially pointless.

This is not to say that dating services – and online culture in general – are not having an effect on our relationships. But what they are mostly doing, apart from helping people find the right match more quickly, is winnowing out those partnerships which are too weak to survive.

As Dan Slater writes in *Love in the Time of Algorithms*, researchers have found that there are three factors that govern one's commitment to a relationship: how satisfied you are with it; the amount of time, effort and resources you have already invested; and the quality of the alternatives available.[91]

This third factor is what, in many ways, is driving the hook-up culture. With so many fish in the sea, it does not become worth the effort to invest in one particular relationship if all you are after is sex. 'Apps like Tinder and OKCupid give people the impression that there are thousands or millions of potential mates out there,' says David Buss, a psychology professor who spoke to Sales. 'One dimension of this is the impact it has on men's psychology. When there is a surplus of women, or a perceived surplus of women, the whole mating system tends to shift towards short-term dating. Marriages become unstable. Divorces increase.'[92]

But if this is happening, it is not borne out by the data.[93] In a good marriage or relationship, the mutual benefits – and sunk costs – will still easily outweigh the temptation to play the field. In a mediocre or failing relationship, the balance may be tilted. But is that really such a bad thing? In the old days, a couple whose spark had gone might have stayed together, for want of a better option. Today, that better option may be just a click of the mouse away – and they may well be happier for making the switch.

True, there is the risk that we become a society of comparison shoppers, forever rejecting adequate partners in pursuit of a mythical perfect match. But the mental, physical and emotional benefits of being in a good relationship are so overwhelming as to provide a powerful glue. In a survey from 2007, 92 per cent of those who married after meeting on the internet said their marriages were perfectly happy, thank you very much – and most said their relationships were stronger rather than weaker for having begun online.[94] Similarly, when the Pew Research Center asked American adults about the effect of the internet on their relationships, three-quarters said it had helped them, and just a fifth that it had hindered them.[95]

THE AUTOMATION OF ADORATION

The great acceleration certainly poses challenges to our relationships – and our sex lives. In *The New Naked*, Dr Harry Fish argues that what he calls 'generation short-sexed' have been trained by internet pornography, and by the rush of life in general, to speed through things in the bedroom.[96] He claims 45 per cent of men climax in under two minutes; as a result, women only get to orgasm 63 per cent of the time. 'We joke that we've "trained ourselves" to be faster at sex in our six years together,' 'Katie' told *Grazia* magazine, in a feature on Fish's work. 'But deep down I know it's not funny . . . if I'm honest I just want to get sex over with quickly, so I can get to sleep or back to my box set.'[97]

Of course, there are all sorts of ways in which technology strengthens, rather than weakens, long-term relationships. For one thing, it helps us find those who share our passions or perversions, however rare. For another, it can help to cement couples' togetherness.

Korea is one of the most wired countries on earth, with its capital of Seoul serving, in the words of Lauren Collins of the *New Yorker*, as a 'sort of terrestrial embassy for the virtual universe'.[98] There, instant messaging long ago displaced phone calls or text messages, and groceries can automatically be ordered by scanning QR codes from a poster on the Tube. And one of the most popular apps, 'Between', can only be used by a couple to communicate with each other: the notes, photographs and letters they send act both as a chronicle of their love and a digital keepsake box, reminding them of their shared past. The service also acts as a sort of personal trainer, enabling couples to set particular emotional goals and cheering them on when they reach them. So popular is the app, says Collins, that it has been downloaded and used by more than half of Korean twentysomethings. They may not have found true love – but they have certainly had all the help they can.

There is also the possibility that technology may actually make us better at romance – or at least more efficient at it. For example, biometric sensors are increasingly being incorporated into our mobile phones. As the economist Tyler Cowen has suggested, it is not hard to imagine an app that causes your iPhone to vibrate every time your conversational partner tells a lie – or, indeed, that can monitor the biological signals from your date, such as pupil dilation, and send you an imperceptible signal when the time is right to swoop in for a kiss.[99] Then there is Tinderbox, an automated plug-in to Tinder that learns your preferences before using facial recognition software to swipe right or left on your behalf – and even initiates conversations in order to determine your intended's interest (though how she will react when she finds out is another story – as is what happens when such bots are common enough that they start checking each other out).[100]

On the less creepy side, similar machine analysis of conversational interactions, or of a couple's genetic information, could help us work out whether a relationship has a sustainable future – or, alternatively, what needs to be worked on to improve it. This may sound heartless – but using technology to foster and strengthen human relationships will surely result in their becoming more durable and enjoyable (especially if the 'Lovetron 9000' catches on – a pelvic implant developed by Rich Lee, a 32-year-old biohacker, which adds a vibrate function to the male penis, complete with multiple speeds and modes).[101]

But what happens when our dependence on technology doesn't merely sit alongside our love life, but comes to supplant it? In Japan, for example, techno-savvy young geeks often hide away from real life in favour of their screens (provoking fears of a demographic crunch due to increasingly widespread sexual apathy among the younger generation).[102] Similarly, a study published in the journal of The Institute for the Study of Labor in Germany found that a decline in the proportion of young men getting married might partly be caused by the ready availability of internet porn (use of religious websites, in the same study, was highly correlated with marriage).[103]

This is the terrain explored by Sherry Turkle. From studying children's behaviour, she has observed how easily they transfer their attentions and affections on to robotic toys, even though these are still fairly primitive.[104] Her worry is that children raised in this manner may never learn to make the effort to reach out to other people, or be able to summon the courage to do so. Instead, they will form relationships with their robots.

This may sound outlandish, but if Japanese people can become so attached to their sex dolls that they give these 'Dutch wives' proper funerals, how much more attached will we be to automatons that can actually provide a genuine facsimile of affection?[105] It is easy to imagine an AI that can iterate emotionally, monitoring feedback from its conversations with you – your smiles, your raised blood pressure – until it has made itself into what is empirically your perfect partner. And who wouldn't then choose the cosy but hollow comfort of such a relationship, given how messy it is to navigate the business of procuring and adapting to a selfish, imperfect human other half?

This is the scenario predicted by the British technologist David Levy in his book *Love and Sex with Robots*: that, as AI develops, robotic partners and helpmates could become widespread.[106] Yet Levy has a much more optimistic take than Turkle (who appears to regard him as something of a Dr Frankenstein). There are millions of people out there, he says, who are being ill-served by the current romantic or even social marketplace, and this trend is only getting worse as the increasing demands of the accelerated workplace strip away the time to develop and cultivate friendships. Levy asserts that Turkle and others:

> miss what I consider to be the main point in developing robots for sex and for strong emotional relationships, which is that there are a huge number of people in this world who are unable to form good human relationships. Millions and millions of people who can't find anyone to love, or anyone to love them.
>
> For these people, life is tragedy. Many, many people have asked me: 'Isn't it much better to have a relationship with a human being rather than a relationship with a computer?' But in my view, that's not the right question. It's: 'Is it better to have a relationship with a robot than no relationship at all?'[107]

Even those who are happily coupled up could, in Levy's view, benefit from a little help from robots. Their capability to learn and provide feedback – and their failure to judge and condemn as humans often do – will make them the perfect sexual tutors, able to help those with particular psychosexual hangups, or who are seeking to define and determine their own sexuality.

'The capability of robots to teach all known aspects of sexual technique will turn receptive students into virtuoso lovers,' he enthuses. 'No longer will a partner in a human relationship need to suffer from lousy sex, mediocre sex, or anything less than great sex.' This will, he says, boost people's confidence, increase their emotional stability – and, yes, save their marriages. On the other hand, it might also lead them, as Turkle fears, to cocoon themselves in their apartments with their robots. As ever, it will depend on individual cultures and people, and how they use or abuse the technology around them.

NO SUCH THING AS SOCIETY?

The speeding up of our social lives – in effect, the transition from offline to online, or rather from offline to a fusion of the two – has proved uniquely suited to scare stories, largely because those lives are no longer being lived just in the real world, but on new platforms hidden from the gaze, or beyond the understanding, of those of a previous generation. Yet for every minus, there is an equivalent plus.

For example, in his famous book *Bowling Alone*, the sociologist Robert Putnam warned that America was fragmenting and atomising, as people retreated into their homes and behind their screens. This had disastrous effects, not just social but physiological. 'As a rough rule of thumb,' wrote Putnam, 'if you belong to no groups but decide to join one, you cut your risk of dying over the next year *in half*. If you smoke and belong to no groups, it's a toss-up statistically whether you should stop smoking or start joining.'[108] But people simply weren't signing up.

Except that they are. What Putnam was looking at was, crucially, the impact of that now familiar foe, television. But when the British writer Henry Hemming, for his book *Together*, attempted to survey as many voluntary bodies in the UK as he possibly could – ranging from beekeeping associations to Rotary Clubs to yoga groups – he found that only 8 per cent had seen their numbers fall over the first decade of the 21st century; the membership of the overwhelming majority was steady or rising.[109] And the reason for this surge? The internet – in particular, the group emails, which had not only made it easier to facilitate face-to-face meetings, but had strengthened and enriched group bonds outside of them.

What kind of people, in the end, is the great acceleration making us? We are harder workers, shorter sleepers and faster thinkers. We expect the world to offer us what we want, when we want it: whether that be using our phones to order takeaways to our door, being able to meet up with our friends, or expecting our job to provide satisfaction and fulfilment. We are more easily upset, or impatient and frustrated, when things do not go our way.

Yet so much of the gloomy portrait that is often painted merely consists of the entirely natural fretting of one generation about the prospects and practices of the next – the idea that, because people are behaving differently, it must necessarily be for the worse. Indeed, we tend to find such worries convincing because that is how our brains are wired: as Clive Thompson has pointed out, studies show that we tend to be more convinced by, and ascribe greater wisdom to, cynical and pessimistic pronouncements than optimistic ones.[110] (This bias also applies, of course, to the prophecies of doom by techno-stress that we encountered in the previous chapter.)

At its most basic, the argument over our accelerating social lives is an argument about whether we will end up using our devices, or being used by them. And which side you fall on will probably be determined not by the results of studies and analyses, but by something more basic: whether you are optimistic or pessimistic about human nature. The argument of this book is that we are not, in the end, rats in Skinner boxes, constantly pumping ourselves with pleasure. Yes, the temptations of technology are powerful – but we are, by and large, capable of turning them to our own advantage.

4

The Art of Acceleration

'We used to wait for it / We used to wait for it / Now we're screaming "Sing the chorus again".'

– Arcade Fire, 'We Used To Wait'

In November 2004, Philip Roth gave a rare interview to *PBS Newshour*. Even though he would not announce his retirement for another decade, the revered author still seemed deeply pessimistic about the future of the artform to which he had devoted his life: the novel. 'I don't think in 20 or 25 years people will read these things at all,' he said. 'There are other things for people to do, other ways for them to be occupied, other ways for them to be imaginatively engaged, that I think are probably far more compelling.'[1]

This is the worst fear of the Slow Reading Movement: that in a more frenetic and distracted age, people will not have the time for great thoughts and great works. Worse, the fact that each form of media must now compete with every other for attention means that they will be driven to get louder, flashier and grabbier. An age of overwhelming cultural abundance will also be one of overwhelming superficiality.

Yet the world of popular culture – the books we read, the shows we watch, the games we play – also offers the clearest possible proof that such concerns are misplaced. In fact, neurological, artistic and economic imperatives are driving popular culture in two very different directions.

One is towards an embrace of the flashy and the trivial – or depending on your point of view, towards perfectly formed and perfectly pleasing micro-experiences. But the other is towards longer and deeper immersion in products of huge sophistication and complexity. This is not so much a reaction to the great acceleration as a complement to it: just as a weekend away or a yoga session helps us cope with the faster pace of the working week, so does sinking into a box set or a 120,000-word bestseller. And this process, alongside broader technological changes, is delivering us more high-quality entertainment than ever before in human history, to the point where we are utterly, ludicrously, spoilt for choice.

KINGS OF CONVENIENCE

At the root of our cultural habits are the neural imperatives discussed in the previous chapters: our cravings for greater convenience, and for more information. We saw earlier how the first of these has serially disrupted the music business, as piracy killed the old model before itself being disrupted by iTunes et al. This is the logic of the instant-gratification economy: consumers will reward services that offer what they want right now, and punish those that don't. For example, iTunes and its cousins are now themselves being disrupted by the instant, on-demand streaming of Spotify, YouTube, SoundCloud and Apple Music.

As for information, we can't get enough, quickly enough. Start-ups are springing up that offer to use 'video summarisation' technology to condense TV shows to their bare bones. Meanwhile, apps such as Spritz or Squirtio promise to revolutionise reading itself by showing you a stream of letters rather than having your eyes scan the page: this, apparently, removes the eye movement that accounts for 80 per cent of reading time, enabling a doubling of reading speed without any loss of comprehension.[2] (Although its long-term success may be jeopardised by the fact that it feels like your eyeballs are being sandblasted.)

This craving for information also helps to explain why, in this new landscape, certain kinds of media are doing better than others. In the battle for attention between the various forms of distraction, the advantage will tend to go to those that are most attention-grabbing, or can be accessed most immediately – or are happy to share our attention. This last factor helps to explain the stunning success of radio in retaining and even building its audience, as well as the powerful new 'multi-screen' symbiosis between television and Twitter, in which each enriches the other. (It also helps if your particular artform possesses the ability to instantly iterate in response to audience feedback – something that marks out, say, video games.)

There is no denying that this appetite for information has imparted a certain breathlessness to popular culture – that just like everything else, the arts are getting faster. Most TV programmes made in the early days of television now seem unwatchably lugubrious: track down an old episode of *All Creatures Great and Small* on YouTube if you want an example. It's the same with film: when *Jurassic Park* was remastered in 3D in 2013 to mark its 20-year anniversary, the *Telegraph* critic David Gritten noted that the most remarkable thing about it was that it took almost an hour for actual danger to appear. 'No Hollywood executive today would feel comfortable with a film that let its well-judged story unfold at such a leisurely, confident pace,' he lamented.[3]

Indeed, if you are prone to despair about the state of popular culture, the state of the multiplex is a good place to start. In an opinion piece proclaiming the 'death of adulthood' in modern culture, A. O. Scott, the chief film critic of the *New York Times*, wrote:

> I have watched over the past 15 years as the studios committed their vast financial and imaginative resources to the cultivation of

> franchises . . . that advance an essentially juvenile vision of the world. Comic-book movies, family-friendly animated adventures, tales of adolescent heroism and comedies of arrested development do not only make up the commercial center of 21st-century Hollywood. They are its artistic heart.[4]

Today's blockbusters are not only superficial in their narrative. Visually, they are dominated by the 'impact aesthetic' – under which you shock and awe your audience into paying attention, then keep it via a relentless sensory bombardment. The paradox is that however hard it hits, this sensory assault often leaves you completely unmarked. As the film critic Jonathan Romney says, 'there's a huge disparity between the time and care and effort that go into these films and how long they linger in your memory. They're like popcorn: they explode with a bang and are completely forgotten almost as soon as you've left the cinema.'[5]

Like today's TV shows, modern films are packed with quicker edits and cuts: the average length of a shot has declined from around ten seconds in the Thirties and Forties to less than four seconds now.[6] There is also more motion and movement within the shot. Both of these work to capture and keep the viewer's attention. Indeed, it turns out that Hollywood has slowly learned to tune its editing to our restless thoughts: scientists have discovered that the fluctuations of shot length within modern films unconsciously match the frequency on which the human brain operates.[7]

The same quickening effect can be seen in music, too. A few years ago, Glenn McDonald of the Echo Nest, a data consultancy later bought by the streaming service Spotify, broke down the details of the 5,000 most popular songs each year between 1960 and 2013, classifying them according to 12 key attributes ranging from 'acousticness' to 'flatness'.[8]

Intriguingly, he found that, while the basic speed of songs has risen, it has not done so consistently: an average tempo of 100 beats per minute rose to 110 by the start of the Eighties, but then dipped back; it is only now approaching a peak again. Similarly, the 'danceability' of the most popular hits has changed little between the days of the Twist and the twerk.[9]

What has risen relentlessly, however, is both the innate loudness of the tracks and their 'energy' – a characteristic defined by 'loudness, beats, structural changes and sounds of the instruments'.[10] In other words, to compete with all the other distractions around us, pop music has got steadily more attention-grabbing.

As well as making songs louder, this process has seen them grow more sonically arresting. The traditional habit of having a slower, quieter passage midway through the song has been cut back – as have the slow-burn intros that might provoke an impatient iPod-wielding listener to shuffle through to the next song.

Even the old cycle of verse-chorus-verse has given way to something more varied. A few years ago, it became common to bring in guest rappers to add a few verses to a pop track. Today, as genres blend into one another, it is more and more unusual to have a song that features just one person, as though producers are afraid that the public cannot bear to listen to a single voice singing similar verses for a full three or four minutes. On a random week in September 2014, the Billboard Top 100 singles chart actually contained 149 separate artists, once all the duets and 'featurings' were included; on any given week that summer, there were usually at least 15 artists squeezed into the top ten tracks.[11]

And what are they singing about? Mostly, these songs are – appropriately – celebrating instant gratification. Today's dance-driven hits generally have a hedonistic, live-for-the-moment vibe: think of floor-fillers such as 'Last Friday Night (T.G.I.F.)' or 'I Gotta Feeling' or 'Shake It Off'. Surveying a list of the top 100 downloads of all time, Gritten's colleague Neil McCormick found it full of 'relentlessly bright and cheery, pick-me-up tracks celebrating love, lust and happiness on the dancefloor'.[12] The role of music today, he added, is not to sit at the front and centre of popular culture, but to soundtrack 'a non-stop, feel-good dance party'.

Yet as with Hollywood's blockbusters, the more mindless these songs appear, the more care and attention has gone into them. One leading producer, Lukas Gottwald, aka Dr Luke, wrote or co-wrote many of the greatest pop anthems in recent years: 'Wrecking Ball', 'Roar', 'Since U Been Gone', 'I Kissed a Girl', 'Right Round'. A *New Yorker* profile described his studio as a modern equivalent of Tin Pan Alley, with a team of more than 40 songwriters at his beck and call – 'artists, producers, top-liners, beat-makers, melody people and just lyric people'.[13]

To produce the perfect slice of disposable three-minute pop, such teams will slave away for months. Each song is stitched together from a variety of different takes, syllable by syllable, to ensure that only the purest and brightest notes come through – before being run through software that punches up every sound so the song stands out in a crowded mall.

The attention paid to such singles is a logical outgrowth of another consequence of the great acceleration – the collapse of the album. In the UK, singles sales have been rising steadily for the past five years, even as album sales have shrunk by approximately 40 per cent.[14] It turned out that once iTunes or Spotify allowed us to cherry-pick the best songs from the 12 tracks on offer, we weren't all that bothered about the filler material. In January 2015, the US music commentator Bob Lefsetz studied a typical week on Spotify: he noticed that the lead single from Fall Out Boy's number one album had been streamed 50,727,320 times, against 5,877,009 and 2,610,943 streams for two of the non-single tracks. 'It turns out people are drawn to the hit,' he wrote. 'Only the hardcore fan listens to more.'[15]

As a result of this process of 'unbundling', says the industry analyst Mark Mulligan, 'playlists and individual tracks have become the dominant consumption paradigm'.[16] There are, he points out, 1.4 million albums on Spotify – but more than a thousand times as many playlists, at 1.5 billion. 'People are spending less time with individual artists and albums. In the on-demand age with effectively limitless supply they flit from here to there, consuming more individual artists in a single playlist than an average music fan would have bought albums by in an entire year in the CD era.' Of course, there are still the hardcore fans who will snap up physical artefacts such as albums (and compilations of outtakes and B-sides), but they are far fewer in number.

This is, inevitably, changing the nature of music, in ways that many traditionalists find uncomfortable. 'There's become a very entrenched mindset in the music industry that the album is the only artform that counts,' says George Ergatoudis, until recently head of music at BBC Radio 1 and 1xtra (which made him one of the most powerful men in the British recording industry):

> I'm not telling artists to stop making albums. If they've got a brilliant album in them, they should go and do it. The problem is that the vast majority aren't bloody good enough – but they may have two or three songs that are incredible.
>
> I honestly don't think that, for the vast majority of people, new albums are going to serve a purpose five years from now. Around ten a year, say, will be so good and so talked-about that actually a lot of people will spend their time listening to the whole album. But even then, most people will encounter it after it's broken into songs within playlists.[17]

SINKING DEEPER

The great acceleration is making popular culture – music, for example – more instant, mindless and disposable. But then how do you account for the fact that the most successful musician of the past decade, Adele, is a slow, soulful, velvet-voiced singer who could have been transplanted straight from the Fifties?

This is because of the crucial point about this accelerated culture: it is not monolithic. There is still space for the kind of songs, or music, that you sink back into. Indeed, given the pressure of our working lives, there is more demand for this kind of product than ever. Again, it's a polarisation at work between fast and slow, with the mushy middle losing out.

You can see this kind of bifurcation in how we behave on Twitter. Watching live shows like *The X Factor* or *The Great British Bake-Off*, or even classy nonsense like *Downton Abbey*, is a fast-paced 'multi-screen' experience: people heckle the show as it goes along, whether in sympathy, outrage or sarcastic glee. Yet for a more complex, engrossing show like *Game of Thrones*, Twitter stays relatively quiet: the flood of posts comes at the end, as transfixed viewers rush online to discuss the latest plot twists.

The same pattern can be seen in all manner of other areas. We complain about how flashy Hollywood movies are – yet the same auteurs who cannot get highbrow arthouse movies made are finding a welcoming home on prestige cable TV channels such as HBO. In books, the biggest sellers in Britain over the past 40 years have often been chunky or challenging reads: *Wild Swans*, *Angela's Ashes* and (the most popular of them all) *A Brief History of Time*.[18]

In fact, perhaps in reaction to an accelerated age, fiction bestsellers appear to be getting longer, rather than shorter. Between 1995 and 2005, the average length of books on the *New York Times*'s sales chart grew by more than 100 pages – and it was already at a hefty 385.[19] The subsequent popularity of such back-breakers as Donna Tartt's 771-page *The Goldfinch*, Jonathan Franzen's *Freedom*, Eleanor Cattan's *The Luminaries* and others suggests that this trend has not gone away in the years since: recently, an article on Slate suggested that there are so many 'megabooks' in the fiction market that we have officially entered the era of the Very Long Novel.[20]

It is, unfortunately, impossible to get empirical proof of this: because Amazon keeps its sales data proprietary, and because it is such a large player in the market, there is no authoritative list of exactly what people are buying. But Mark Coker, the founder of the independent eBook distributor Smashwords, has certainly found that sales tend to increase along with word counts. The top ten titles on his site came in at just over 100,000 words on average – which is far longer than most traditional printed novels. The top 100 averaged a shade under 100,000, and the top 1,000 titles around 73,000.[21]

When you consider that the average length of a book in the Smashwords database is just 37,000, it seems clear, as Coker claims, that 'readers are going out of their way to search out and purchase longer eBooks'.[22] It is only in genres explicitly marketed for their disposability, such as romance and erotica, that his bestselling titles tend to come in at under 100,000 words. There is a parallel here with video games – long considered the home of mindless, point-and-shoot gratification – where the biggest titles actually market themselves on how long they take to complete, with 60 hours relatively standard.

And there is yet another piece of good news about the state of the popular culture. It is that whether they be fast and poppy or slow and highbrow, the products we are consuming are becoming more complex, and more rewarding.

This is a big claim to make. But in his marvellous book *Everything Bad is Good for You*, Steven Johnson makes the point that it is not just the pace of TV shows and the like that has increased, but their information density – and their level of intellectual challenge.[23] If you study an episode of *Dragnet*, says Johnson, then *Hill Street Blues*, then a more modern show like *CSI*, you will find that the number of characters, relationships and plot points increases dramatically. In other words, we cannot be bombarded with information without developing an enhanced capacity to cope. That is why *The Wire* or

Breaking Bad really are better, or at least more sophisticated, than anything that has come before them: because they have an audience that can keep up.

Some creators even marry the depth and complexity of long-form with the speed of short-form – a fusion that can be offputting for some, but dizzyingly compelling for others. The latest big thing in US TV is series, such as *Empire* or *Scandal*, that burn off plot at such a pace as to keep their viewers hooked by the rollercoaster ride.[24] And then there is the British television writer Steven Moffat. The producer of *Doctor Who* and co-creator of *Sherlock* is known for the complexity, density and rapidity of his plots, leaving huge gaps of logic and chronology in his stories and challenging the audience to keep up. As the late Clarissa Tan wrote in the *Spectator*:

> [*Sherlock*] is no longer about the detective outwitting the criminal, but about the programme outwitting us. Time and again . . . we see timelines being spliced or fast-forwarded, so that we the viewers are left in no doubt we are but putty in scriptwriter Steven Moffat's hands . . . then there are the plot twists that come at such a furious, dizzying pace that you can't help but admire the show and be hooked on it . . . Whether or not you think *Sherlock* is good drama, this twist-every-two-ticks approach changes our expectations of television. It's like a dopamine hit, the same kind that makes us keep checking Twitter or Instagram or whatever, and after this we can't go back.[25]

Tan had a point: there is something about Moffat's writing that seems to embody the very essence of modernity. Not least because, when challenged about his plot's pacing and complexity by *Doctor Who* fans and critics who struggled to cope, his defence was that his young audience understood perfectly well, because they were able to cope with a volume and flow of information that their elders couldn't handle.[26]

As consumers, all of this leaves us better placed than ever. If we want mindless distraction, the examples on offer are lavishly and impeccably produced; if we want sophistication, there is Radiohead, or HBO, or even Philip Roth. The result is that, in many respects, the great acceleration has led to a genuine cultural golden age.

THE SHOCK OF THE NEW

It is at this point that some people will raise a significant objection. How can we call it a golden age – still one of ceaseless, restless innovation – when so much of pop culture seems devoted to pursuing not things that are new and different, but more of the same? Pop songs are packed with attention-grabbing hooks and vocal changes, but they still feel like they came off the same conveyor belt. Multiplexes are filled with sequel after sequel, while the video-game calendar is fixed around the release of updated versions of billion-dollar franchises such as *Grand Theft Auto*, *Call of Duty*, *Halo*, *Battlefield*, *Assassin's Creed*, and so on. And if it's not sequels, it's clones: every surprise success is followed by a host of me-too titles.

Yet this phenomenon, too, is a by-product of acceleration. The reasons for that are both biological and economic. Our brains are, as we have seen, programmed to appreciate novelty. But they also like familiarity and repetition: the flow of dopamine along the same pre-etched channels. So the key to hitting the neurological sweet spot is to offer us something that sounds or looks close to what we know, but is different and novel enough to challenge our expectations.

As the neuroscientist Daniel Levitin writes in *This Is Your Brain on Music*, the secret to a really great tune is that it establishes expectations and then toys with them.[27] David Brooks, citing Levitin's work, gives the octave leap in 'Somewhere Over the Rainbow' as his favourite example. 'Life is change,' says Brooks. 'And the happy life is a series of gentle, stimulating, melodic changes.'[28]

This pattern is mirrored in how we watch television: we like novelty, but not too much. In sitcoms, for example, gags now come thicker and faster than they once did: the US critic Todd VanDerWerff has charted how shows like *Seinfeld* and *The Simpsons* accustomed Americans to a quicker pace of humour.[29] These were followed by a wave of successors like *Community*, *30 Rock* and *Arrested Development* that took this to an extreme, racking up huge joke-per-minute counts. Yet as VanDerWerff says, while these shows all won passionate cult followings, they never became mainstream hits, in large part because they were moving too fast for many people's comfort. In both Britain and America, the true comedy juggernauts – *Two and a Half Men*, *The Big Bang Theory*, *My Family*, *Mrs Brown's Boys* – deploy their gags at a more comfortable pace, even if it is still faster than in the old days.

Perhaps the best example, however, is the story of OutKast's monster hit single 'Hey Ya!', told in Charles Duhigg's book *The Power of Habit*.[30] Within the music industry, there is a Spanish company called Polyphonic HMI – one of the new breed of big-data outfits that seek to apply algorithmic certainty to the messy business of creativity. Its software made its name by flagging up Norah Jones's debut album *Come Away With Me* as a hit in waiting, even though she was a complete unknown.

When executives at the record label Arista put 'Hey Ya!' through the same software, the results were incredible: it got the highest scores ever seen. The bosses knew they had a monster hit on their hands: even without the software, it was easy to tell the song was relentlessly, astonishingly catchy. But when they started playing it on the radio, it bombed.

As Duhigg says, OutKast's problem was that their song was simply too original. 'It didn't sound like other songs, and so some people went nuts when it came out,' recalled John Garabedian, a leading US DJ. 'One guy told me it was the worst thing he had ever heard.'[31] Whenever the Philadelphia radio station WIOQ started playing it, 26.6 per cent of listeners actually changed channel.[32]

Still, the station's bosses really liked the song – and the data clearly showed that it had huge potential. So they started using what was called 'sandwich' programming: they would play 'Hey Ya!', but put it between two other songs that audiences were proven to keep listening to because they were comforting and familiar, even if they didn't actually like them that much – a Celine Dion track, say. Soon, the proportion of listeners turning off when 'Hey Ya!' came on started tumbling – because it had become associated with the familiar and comfortable. A few months after its controversial debut, WIOQ was pumping the song out 15 times a day.[33]

It is not just about neurology. The other reason why so much mainstream culture has a familiar feel (and where the great acceleration comes in) is brute economics – in particular, the way that entertainment, like so many other fields, has become a winner-takes-all affair.

A few years ago, this would have seemed a strange claim to make. The vision most frequently set out – most famously by Chris Anderson, editor of *Wired* and author of *The Long Tail* – was that culture would not so much decay as divide.[34] The idea of a 'mainstream' would be abandoned as a flood of niche programming fragmented the audience. Farewell to the nation coming together to find out who shot J. R. and hello to a million teenagers broadcasting from their bedrooms.

The problem is that, in financial terms, this doesn't actually seem to be happening. What we see instead is the same tendency towards gigantism that we observe in Silicon Valley. Yes, plenty of niche content is indeed being produced. But as the Harvard business professor Anita Elberse has convincingly demonstrated, the hits appear to be getting bigger and the tail appears to be getting smaller.[35]

In 2011, for example, roughly 8 million singles sold at least one copy in the US. But 7.5 million of them – 94 per cent of the total – sold fewer than 100 units.[36] That's not a business model: that's vanity publishing. And of the songs that did sell, the big hits sold in staggeringly larger proportions: the top 102 songs, just 0.001 per cent of the total, accounted for 15 per cent of the market. Their dominance becomes even more pronounced when you realise that many of the smaller sellers within the long tail are former chart-toppers, enjoying a lucrative afterlife.[37]

Contrary to Anderson's theory, this pattern turns out to be getting stronger rather than weaker. In 2013 the ten biggest songs of the year were played on US radio twice as much as the equivalent songs a decade before: Robin Thicke's 'Blurred Lines', the year's biggest hit, was played 70 per cent more than its 2003 counterpart, 3 Doors Down's 'When I'm Gone'.[38]

Hollywood has also been castigated for its reliance on blockbusters and sequels, but it too has good economic reason to cling to them. Warner Bros calculated in 2010 that its three top movies had cost 30 per cent of its budget, but made more than half of its international profits.[39] Today Disney, the most successful modern studio, isn't spreading its bets on an array of niche content: it's investing billions to make more *Star Wars*, more Marvel, more Pixar, more *Frozen*, which can then be cross-marketed across its vast empire of TV networks, theme parks and toy stores. And in TV, when firms such as Netflix and Amazon scramble for video-streaming subscribers – as part of their own attempt to disrupt the original networks – they are throwing their production and marketing budgets behind not hungry young unknowns, but original series from big names such as David Fincher and Kevin Spacey, in the shape of *House of Cards*, or Garry Trudeau and John Goodman.

Anderson, it turned out, was right about the great acceleration breaking down barriers to entry. But he had failed to take into account the higher barriers to success – not least because there is so much amazing content out there that breaking out of the crowd and capturing people's fleeting attention is harder than ever. More than half of new TV shows now don't see a second season, compared with just 10 per cent in 2000 – which is also a symptom of increased impatience among both audiences and executives.[40] To cut through the clutter, you have to produce something so good that it stands out on its own terms. But you can also trade on existing brand recognition – in other words, make yet another superhero movie.

FOLLOWING THE CROWD

This blockbuster model also thrives because of another trick of our brains: the extent to which we are social creatures. This is evident in the renewed appeal of live spectacles on television: events such as *The X Factor* or the Super Bowl that can be tweeted as well as watched. But it also manifests itself in our desire to go along with the herd.

In *Blockbusters*, Elberse shows how this works. A group of scientists, she reports, set up their own version of iTunes, which let people see how many times a song had been downloaded. As you'd expect, the most popular tracks tended to do better. But then, the researchers reversed the order – they pretended to the site's visitors that the most popular songs were the least popular, and vice versa.

It turned out that people were more willing to trust the crowd than the evidence of their own ears – indeed, they convinced themselves to share the group's opinion. As a result, while the very best songs always made it back up to the top eventually, and the very worst sank to the bottom, the placement of most of the others was essentially random.[41]

What this means, says Elberse, is that, yes, the cream will rise to the top. But otherwise, 'the ultimate success of an entertainment product . . . is extremely sensitive to the decisions of a few early-arriving individuals'.[42] One good review, or positive piece of buzz, can create a tiny advantage for a particular film or song, which becomes hugely magnified as later buyers pile in behind the early adopters.

The only logical choice for producers and executives, therefore, is to do all they can to make their product launches as big and splashy as possible, in the hope that they will gain that early audience. Alan Horn, who pioneered the blockbuster strategy at Warner Bros, told Elberse that 'the average moviegoer in the US only sees five or six movies a year. And it is even fewer in international territories. [In 2011], there were over 120 films released by the six major studies, and another 80 by the larger independents . . . that is why having something compelling is so important – something of high production value, be it because of the story, or the stars involved, or the special visual effects.'[43]

Making such a splash becomes all the more important as our entertainment options approach the infinite, and capturing eyeballs and standing out from the crowd becomes ever more difficult. And adding to the pressure to secure instant success is the fact that, in such a crowded marketplace, the window in which a product can find success is becoming ever shorter. In 2011, the top 100 films at the US box office collected 30 per cent of their takings in their first week alone.[44] In publishing, the proportion of a book's lifetime sales made in its first month on the shelves has risen from 30 per cent to 42 per cent, according to one marketing director.[45] Shorter public and corporate attention spans mean that if a book hasn't broken through within six weeks, it is effectively relegated to the backlist.[46]

And this process, too, is speeding up. Hollywood has long lived and died by the opening weekend – but now, in the age of Twitter, films can be declared a stinker or a masterpiece within moments of the first critics coming out of the first screening. This process is accelerated by the pressure from news desks to dash off the first definitive review, and thereby capture as much of the ensuing search traffic as possible – and to make a splash by offering the strongest opinion possible. Some editors have even been known to ban their critics from filing three-star reviews, because no one wants to read anything that doesn't give them a jolt.[47] In books, the equivalent is the Amazon bestseller list: getting near the top of it is not just a quality signal to the reading public, but tells the big buyers – such as supermarkets and retail chains – that they need to stock up, fuelling the title's success further.

The growing dependence on blockbusters doesn't just spring from the nature of the market. It is also driven by the nature of the companies making these products. Most modern corporations, as chapter 7 will make clear, are obsessed by short-term growth – in particular, by hitting their quarterly targets to keep shareholders happy. But what happens when the market is mature, and growth is harder to come by? There are, in the West at least, pretty much as many people buying books or going to films as there are ever likely to be. Yet still the pressure comes down from on high for bigger profits, year after year. If executives fail to deliver, they are, like football managers, pushed out.

As one of those executives, then, your only options are to buy growth – by acquiring other companies – or to increase your revenues. And when you're a big company, the only way to do that in the short term is to make the kind of big bets that are likely to shift the financial needle.

In *Merchants of Culture*, his study of the modern publishing industry, John B. Thompson describes the resulting rise of what he calls 'extreme publishing' – the buying of books, often by celebrities and often ready-made, that can be shoved into the schedules outside of the normal slow process of commissioning and revising, in order to juice up profits.[48] These are books that are 'easy to define, easy to sell and easy to communicate'.[49] The risk, however, is that they are also criminally undercooked – or that one or more of these big, expensive bets will not come off.

It is a similar need for outsize hits that lies behind Hollywood's reliance on familiar formats and topics – especially superheroes. To make enough money, a blockbuster needs to be a 'four-quadrant' hit, appealing to young and old, men and women. Often, alas, this means catering to the lowest common denominator – particularly given the accompanying need to sell it to the international market, where an explosion always goes down better than a dramatic moment. Indeed, big-budget comedies are, the statistics show, something of a dying form, given the difficulties of finding gags (at least, those not involving bodily functions) that work as well in Shanghai and Seoul as Missouri and Manhattan.[50]

What we are seeing here is not the long tail, but winner-takes-all – or at any rate winner-takes-most. And it's a pattern that repeats across an accelerated world: one of the chief effects of technological speed is that it makes it much easier for successes, whether they be apps or albums, to spread across the globe, and for word of mouth to build behind the most popular. Indeed, in many cases the same giant firms who've benefited from acceleration elsewhere are controlling, or at least influencing, this market too: being on the front page of the iTunes store, or picked out by Amazon, counts for far more than a five-star review in the *New York Times*. It also helps that such dominant firms (for example Amazon in books) can monopolise sales data within their platforms, giving themselves a huge informational advantage.

And, as elsewhere, this process has been accompanied by a hollowing out of the middle. As Alan Horn admitted, focusing his efforts on $200 million movies means that he was unable to make as many $90 million titles – or that he had to cut their budgets down to, say, $60 million.[51] The consequence of that has been to drive even so renowned a cinematic craftsman as Stephen Soderbergh out of Hollywood and into the arms of HBO – despite the huge success of his *Ocean's Eleven* franchise.

WATERING THE GRASS ROOTS

It may seem upsetting – even alarming – that the big beasts of culture are committed to playing it safe. But that ignores the astonishing profusion of quality content that is springing up alongside their work – and often, beneath their feet – and of new channels for that content to find an audience. For example, the fact that quality television series of the kind produced by HBO have become hugely successful – just look at the DVD sales for *Game of Thrones* – accords with Anderson's thesis. What matters to subscription services such as HBO or Netflix or Sky is not raw ratings, but monopolising niches: providing the very best content for fans of sport, or brooding murder mysteries – or Kevin Spacey. That is why cult series such as *Arrested Development* have found a second home online.

But beyond that, there lies the boundless creativity of a planet of active, engaged users – on Twitter, YouTube, Tumblr, or wherever. Today, teenagers are not just broadcasting from their bedrooms, but coding and recording from them, too. Indeed, by any standards, we are living through an explosion of mass creativity.

Thanks to the technological advancement that the great acceleration has brought, the cost of becoming a broadcaster, or a computer games programmer, or a special-effects whizz, has never been cheaper: in computing, 'middleware' programes such as Unity have made it drastically easier to produce a remarkably polished iPad game, just as it is so much easier to set up a start-up company. The ProTools software that many amateur music editors use contains the same capacity as a $1.5 million mixing desk would have done in the Eighties.[52]

Moreover, thanks to that same technological acceleration, these producers can broadcast this to a worldwide audience – ideally finding enough fans to be able to make a living off it. Just consider the success of *Minecraft*, which began as the personal project of an unknown Swedish programmer, Markus 'Notch' Persson, and grew to absorb millions of man-hours (and child-hours) of patient construction: it became such a huge phenomenon that it was eventually bought by Microsoft for $2.5 billion.[53]

Digital distribution and the increasing popularity of funding platforms such as the crowd-funded Kickstarter make it easier than ever for outsiders to put their ideas forward, and to attract an audience and create a market for them. Of course, much of the new material is rubbish: but then, as Sturgeon's Law has it, 90 per cent of everything is crap. The remaining 10 per cent, however, is often really rather good.

Perhaps the best example of this – and of the effect of technological disruption on the media – is the success of YouTube. When we talked about Zelda in the last chapter, we mentioned that she was watching much less TV than her parents. In fact, in 2014 the research analyst Michael Nathanson reported that the average age of American TV viewers was far higher than that of the population as a whole, and getting older much more quickly: the typical viewer of the major network shows is now 53.9 years old, roughly 15 years older than the national average.[54]

So what are teenagers doing instead? In large part, watching YouTube. On 3 August 2013, two American Muslim video bloggers, Adam Saleh and Sheikh Akbar, decided to have a meet-and-greet with their fans at Marble Arch in London – and ended up having to be rescued by police after hundreds of shrieking, hijab-wearing girls appeared to catch a glimpse of their idols.[55]

Saleh and Akbar are not alone. Names and channels such as Zoella, Slomozovo, TyrannosaurusLexxx, TomSka, Michelle Phan, Charlie McDonnell, FoodForLouis, Marcus Butler or Tyler Oakley may not trip off the adult tongue, but mention them to your teenage daughter and her reaction will be rather different. They are part of a new wave of stars created by video blogging – whether it be singing songs, sharing make-up tips, performing stand-up comedy or, in the case of FoodForLouis, eating all manner of disgusting creatures, including live ragworm.[56]

This is starting to become seriously big business. The top five YouTube stars now have more subscribers than the population of Mexico.[57] Phan, a beauty blogger, has a retail empire estimated to bring in $120 million a year.[58] PewDiePie, who broadcasts videos of himself playing video games, earned $7.5 million in 2014.[59] The market has become mature so fast that in May 2015, an article on *Re/Code* could seriously ask 'Can You Still Become YouTube Famous?'[60]

The type of content that populates YouTube, or Snapchat, or Vine, may often seem infantile or unpolished to the adult eye (especially on Vine, with its six-second video limit). Yet there is plenty of wit, humour and intelligence, too. And the growth of such platforms is built on the rise of a generation who are both producers and consumers of content – 'prosumers', as the marketers have it. Even a decade ago, more US teens claimed to be creating content for the internet than consuming it; in a more recent survey, the proportion of self-proclaimed 'creators' had risen to two-thirds.[61] And whether they want to share photos, contribute to webpages, remix videos or produce their own songs, the web has made it drastically easier to do so.

Despite its growing significance, video blogging and the creative ferment it has inspired had until recently mostly gone unnoticed in the mainstream press, largely because the generation that controls the commanding heights of the media prefers to dwell on its own youth rather than other people's. 'Most people treat YouTube in the same way as they would a blocked toilet or Piers Morgan's TV career,' claimed Benjamin Cook, a fuchsia-haired video blogger who profiled other stars of the medium in his 'Becoming YouTube' series, in an interview with the *Observer*. 'They don't know how it happened or who's behind it, but they figure it's probably just full of shit and they'll leave it for someone else to deal with.'[62]

Some, indeed, see these YouTubers as just another example of rampant narcissism: a bunch of kids who all want to be famous, even if they don't quite know for what. But they are actually the children of the great acceleration, not least because their world is relentlessly, remorselessly social. The selling point of these young stars is their accessibility: they are people just like you, in a house just like your own.

Admittedly, there is a tension here: the bigger YouTube gets, the harder it is for stars to retain that personal connection with their fans. At one of the three big YouTuber conventions, London's Summer in the City – 'a sort of Woodstock-cum-Glasto for the selfie generation', according to a *Financial Times* leader writer dragged along by his daughter – there are now rope lines and green rooms to marshal the queues and segregate the stars, who are said to be demanding thousands of dollars (via their new agents) for appearances and endorsements.[63] DigiTour, which takes YouTube idols on the road to enable their largely female fans to shriek their approval, sold 18,000 tickets for its shows in 2013, 120,000 in 2014 and roughly a quarter of a million in 2015.[64]

But for all that, there is still a sense of informality and freshness to the medium – this is a quicker, more direct, more impromptu and more personal form than traditional television, almost closer to theatre in the way its performers have to improvise and build a rapport with their audience. 'If TV is a monologue,' says Benjamin Cook, 'then YouTube is a conversation.'[65]

For many reared on traditional forms of media, this informal, self-starting, instantly reactive environment is deeply uncomfortable. They lament the decline of the middle-ranking, middle-brow artist – the fact that these days you are either a megastar or a niche artist, appealing to everyone or to your own particular fanbase. As John B. Thompson says of publishing, the new ecology delivers outsize rewards to the lucky few, but:

> for the vast majority of writers or aspiring writers, this system seems like an alien beast that behaves in unpredictable and erratic ways,

> sometimes reaching out to them with a warm smile and a handful of cash . . . and then suddenly, without much warning or explanation, pulling back, refusing to respond or cutting off communication completely. This is a system geared towards maximising returns within reasonably short time frames; it is not designed to cultivate literary careers over a lifetime.[66]

So, how to cope with this as an artist? Well, largely by embracing the opportunities that acceleration provides – as those YouTubers have, or the self-published authors on Amazon's Kindle. Elberse's findings may have suggested that most people won't make it – but they also showed that the very best content really does tend to find an audience. And when it does, it tends to succeed explosively.

In this new world, acts have to work not just to build their fanbase, but to keep it. In a *Wall Street Journal* op-ed, Taylor Swift cited the example of an actress friend who was one of two potential choices for a movie role: rather than resorting to the casting couch, the director chose the one with the most Twitter followers.[67] Indeed, performers have to work harder all round. Given the public's limited attention span, it is dangerous to remain out of view for too long – so pop stars such as Rihanna seem to be on a perennial cycle of touring and releasing singles, or at the very least guesting on others' songs.

But is it such a bad thing that acts have to remain on their toes – that it is no longer as easy to put out an album stuffed with filler? Whereas in the old days each artist established their brand and stuck to it, increasingly it is all about whether the song itself is any good. The British singer Ellie Goulding, for example, recently had two songs in the Top 40 at the same time. One was an R&B-inflected number called 'Burn', written and produced by the usual assortment of hitmakers-for-hire; the other, a saccharine, slow-paced cover of a ballad called 'How Long Will I Love You?'. Not only did no one seem to mind the stylistic discrepancy, no one even seemed to notice. David Beer, a sociologist at the University of York, calculated that in 2014, the UK No. 1 song changed 41 times – and 14 separate artists reached No. 1 with their debut single (for context, no act at all managed that between the chart's launch in 1952 and Whigfield's release of 'Saturday Night' in September 1994).[68] It is musical turnover – in terms of both songs and acts – 'at a rate that we've never seen before'.[69]

There is an argument that such an environment isn't exactly great for an artist's long-term development, or earning power. But for the audience, this is outweighed by the short-term gains. 'The volatility in the audience in terms of how quickly they will make a decision that they're no longer into that artist, or that artist isn't meeting their standards, is incredible,' says George Ergatoudis. 'Lady Gaga, after the initial grind, had explosive success and looked like a global phenomenon in every sense of the word. But there's been a complete law of diminishing returns on each of her albums, to the point where you'd have to question how bankable she is.'[70] That isn't her fans' fault: it's her music's.

PLAYING THE GAME

The volume and quality of cultural content being produced – whether snappy and instant or complex and engrossing – makes an irresistible argument that, as consumers, we have never had it so good. But for final proof, consider a cultural sector often neglected by the traditional media, but dear to my heart and that of millions of others: the world of video games.

Games are, as most people now realise, big business. But it's still hard to get your head around how big. In 2014, for example, Amazon paid almost a billion dollars for a firm called Twitch.[71] What Twitch does is very simple: it broadcasts people playing computer games. That might range from an expert sitting in his room talking less experienced players through the trickiest bosses in the *Dark Souls* series (itself a great counter-argument to the idea that the digital age has made us less patient and determined, given the dozens of hours needed to master the game), to friends sharing games of *Grand Theft Auto*, to the broadcasting of sporting events such as *Street Fighter* or *Starcraft* tournaments.

The fact that Twitch became a huge player in the online video market very quickly indeed reflects the fact that we are living in a more social world: we all want to peer over other people's shoulders to see what they're up to. But it also reflects the huge appeal of the genre as a whole.

It is, of course, impossible to easily sum up as sprawling a field as the modern video games industry. Yet the main trends conform exactly to those described in the rest of this chapter. For starters, there has been an explosion of content: new tools such as Unity have made it easier than ever to create games, or content for them. Next, there has again been huge disruption due to technological change. A few years ago, a firm called Zynga was lionised for mastering the new gaming ecosystem of Facebook – but the move to mobile devices (as well as a tweak in Facebook's algorithms to prevent Zynga flooding users' News Feeds with updates about *Mafia Wars* or *Farmville*) left the company stranded. Similarly, Nintendo went within only a year or two from being top of the heap to bottom of the pile, as the mass casual audience that powered the enormous success of its Wii and DS consoles migrated to Apple's devices alongside those Zynga users.

In response to these trends, the industry has fragmented in a familiar way – by splitting into lions and flies. Giant mega-firms have emerged with the resources to fund ever flashier (yet wearily predictable) blockbusters such as *Call of Duty*, as have a host of smaller, nimbler independent firms. The ones who have suffered are, again, the mid-sized publishers, many of whom – like their equivalents in publishing – have been absorbed by the giants.

Yet this is not a David and Goliath situation, where the independent sector stands opposed to the corporate leviathans. On the contrary: despite the occasional flurry of hostility, their relationship is symbiotic, just as Amazon's is with the writers whose works appear on the Kindle, or Apple's is with the people who produce apps for its store.

Microsoft, Sony and Nintendo might sell their consoles primarily with the big 'AAA' titles they have developed themselves or procured from their fellow mega-firms such as Activision or Electronic Arts – but they know that offering a vibrant selection of smaller, cheaper and more experimental titles is vital in making their whole proposition more attractive. Thanks to the rise of digital distribution, the developers of such games can offer them via far more formats and platforms than ever before, keeping the lion's share of the proceeds – and if they can't find the funds to make them, they can turn to their fans via Kickstarter, or the Early Access program on the PC gaming platform Steam. The result has been an explosion of quality titles, many of them tackling complex, adult themes or focusing on the aesthetic or philosophical experience rather than on mindless fun (see, among many others, *Journey*, *Papers Please*, *The Stanley Parable* or *Device 6*).

What we are seeing, in other words, is that familiar segmentation between instant gratification and longer, more in-depth experiences, with niche and cult products sitting comfortably alongside. But what is especially interesting is the way that all of these titles are being engineered more ingeniously than ever before to provide loops of satisfaction stretching across seconds, minutes, hours or even years.

Most of us realise by now that mobile games such as the infamous *Candy Crush* – played 150 billion times in its first year[72] – are designed using a precise range of systems to keep players hooked, with every light and sound and swipe of the screen calculated and focus-grouped to keep you tapping away for just that little bit longer. The mantra of Bungie, developer of the mega-selling *Halo* series and the more recent smash hit *Destiny*, is to offer the player '30 seconds of fun' – an intense, thrilling shoot-out with alien enemies – and then repeat that again and again, with suitable variations, to provide hours of seamless pleasure. Its levels are rigorously play-tested to ensure that there are no awkward pauses, that it is never too hard to find the next battle arena on the map.[73]

On a slower timescale still, there is the 'grind' of massive online role-playing games such as *World of Warcraft*: its designers, like those of *Candy Crush*, can track millions of decisions made by players and alter anything that is causing people to drop out of the game or become frustrated, ensuring a constant smooth dose of dopamine as enemies are vanquished or levels conquered. As the games theorist Edward Castronova says, 'It's like the discovery of a new continent. What we're developing here is a science of how to make people happy, and that's both a really exciting and a dangerous thing . . . [it may be that] being hooked up to an experience machine that makes you happy all the time is not a good life.'[74]

There has, inevitably, been some worry that this may shade into addiction: that we will become trapped in these impeccably designed Skinner boxes. The most notorious case is that of a South Korean couple who let their real baby die because they were too busy caring for one in a computer game.[75] Yet as with cyberbullying and other unpleasant online phenomena, the cause of such behaviour is more often actual than virtual. The great acceleration gives people ways to show they are broken; it does not break them in the first place.

WHO NEEDS HUMANS ANYWAY?

The techniques above – and, indeed, the techniques for evaluating songs used by Polyphonic HMI – raise a final, fascinating question: will we end up with a culture that rewards not artistic originality, or personal creative expression, but calculated attempts to push the right neurological buttons? And might it not all start to seem a little – well, samey?

With Hollywood, as we've seen, it's impossible to watch the latest blockbusters without feeling that you've seen this movie somewhere before: the explosions may be different, but the plot contains the same precisely calculated sequence of successes and reversals, action and introspection, with each character's emotional arc resolving itself with satisfying neatness at a perfectly predetermined juncture. It's as if it's obeying a formula – and in fact it is.

The formula comes from a little-known 2005 book by the late Blake Snyder, a successful screenplay guru, called *Save the Cat! The Last Book on Screenwriting You'll Ever Need.*[76] It breaks down the traditional three-act structure into 15 'story beats', with titles such as 'Catalyst', 'Theme Stated', 'Bad Guys Close In', 'All Is Lost', and prescribes exactly when each one should fall within the film. Once you've read it, all of your doubts and annoyances about the familiarity of, say, *Guardians of the Galaxy* snap into focus: we're not just getting Joseph Campbell's myth of the hero's journey, endlessly retold, but a step-by-step guide to the path that journey will follow, and the precise moment at which the love interest's encouragement will inspire the protagonist to victory.

Snyder's formula – which has since become holy writ among the studios – was based on observation: years of experience of watching and writing films, and working out what hit the spot with audiences. But what happens if we apply to that effort the same kind of relentless, automated focus-grouping seen in computer games – or, indeed, in Silicon Valley?

Big data is now a mainstay of the culture industry – and is being used in increasingly sophisticated ways. There is a range of services which are designed to track the performance of songs (or any other form of media) once they've been released into the wild: Shazam's software lets record executives see the popularity of songs or bands spread like a virus from town to town, Facebook post to Facebook post, as does a service called Next Big Sound (recently acquired by the internet radio company Pandora).

Others, however, are devoting their number-crunching efforts not to how songs or films are received, but to their raw ingredients. Firms such as Netflix or Spotify devote enormous efforts to analysing and categorising films, music, books and so on, in order to tease out fundamental similarities and thereby improve the performance of the matching algorithms that power their playlists and recommendations.

This kind of analysis can be done by analysing either the product itself, or – more ingeniously – what people are saying about it. This is the approach taken by the Echo Nest, now bought up by Spotify, which, as well as analysing the raw musical information, has catalogued how people talk about millions of songs by scrutinising blogs, Facebook and other sources.

The goal, says its 'data alchemist', Glenn McDonald, is to create software that – purely by studying the musical information within a particular song – will be able to play exactly the right follow-up track. Indeed, he rates this as an explicit cure for the distraction fostered by acceleration: 'There are millions of people who love music and care about music, but either don't know how to expand their life or don't have time to. I'm trying to help those people remain engaged in music.'[77]

At the moment, he says, the system works decently if you combine the 'metadata' – the band's name or the genre – with the raw musical information. Rely on the latter alone, however, and the algorithms fail 'spectacularly': American country music might, mathematically, be surprisingly similar to a particular form of Indonesian folk-pop, but devotees of the one wouldn't be pleased to find the other turning up on their playlist.

Such categorisation is a subject of huge commercial interest for firms such as Spotify, as they strive to keep people listening for as long as possible in the face of so many other entertainment options.

But there is another, and even more intriguing, reason to undertake this kind of analysis: to improve the products themselves. Might it be possible to arrive at an algorithmically, mathematically perfect pop song, or action movie one-liner?

Well, people are certainly trying their best. Just as Polyphonic HMI or Music X-Ray claim to be able to tell you whether a song will be a hit, so will a firm called Epagogix feed the details of your upcoming movie into its algorithms and predict your expected profit. But more importantly, Epagogix can tell you what needs to change to boost that profit – some reworking of the third act of the script, the replacement of vampires with zombies as the antagonist, and so on. One of its most interesting findings a few years ago was that who stars in a movie made little difference to the eventual box office, with four exceptions: Will Smith, Brad Pitt, Johnny Depp, and one hapless leading lady who actually dragged takings down.[78]

Such systems are as yet in their infancy: for all Polyphonic HMI's apparent success, Glenn McDonald's own algorithms have yet to find any simple sonic rule of thumb – a particular key or tempo, say – that separates the good songs from the great ones. 'I expected to find that the more energetic the song, the punchier, the bigger a hit it would be in general. But in fact, that doesn't hold. You don't have a better chance of having a hit song just because you've set a particular knob to a particular setting.'[79]

That doesn't mean, however, that these correlations don't exist, it just means that they are elusive. And as the tools to find them evolve, they are surely set to become an inevitable part of every executive's decision-making process. Netflix, for example, did not invest more than $100 million in remaking *House of Cards* because it desperately wanted to work with Kevin Spacey and David Fincher: it did so because its data signalled that existing fans of the original BBC series also tended to be Fincher and Spacey nuts.[80] The Echo Nest's own musical surveys have thrown up similar strange connections, which offer similar opportunities for cross-selling: for instance,

Beatles fans also tend to be into Stephen King, while Lady Gaga fans are also keen on Michael Kors, Ellen DeGeneres and Zooey Deschanel.[81]

It may not be long, given the acceleration in computing power, before similar software can carry out the improvements to a film's script or a song's harmonies on its own: after all, artificial composers have long been capable of producing classical pieces indistinguishable from the work of human beings.

Yes, this may produce some unintended consequences. As mentioned above, such algorithms have no idea what a 'good' song or film actually is: the big-data approach essentially relies on feeding in records of what has and hasn't worked in the past, and drawing comparisons. One danger to our long-term cultural health is that up-and-coming artists – assuming they haven't been driven out of the business by robot competition – will devote themselves to gaming the algorithms rather than developing their own voice. This is a problem explicitly acknowledged by Next Big Sound, which uses big data to track the 'buzz' around various artists: because the major labels monitor its service, bands are incentivised to win its approval rather than the audience's.[82] By observing, the site changes the thing being observed.

A data-driven culture in which performers mimic what has gone before risks privileging craftsmanship over genius, and making it harder for genuinely original performers to make their mark. Andrew Leonard lamented in an article for *Salon*: 'Can the auteur survive in an age when computer algorithms are the ultimate focus group?'[83] Yet human taste has always been changeable and diverse. New artists, genres and ideas have always managed to emerge and challenge existing orthodoxies: the lesson of centuries of creativity is that there is always space for great writers or musicians to move the needle – to innovate and invent and challenge established procedures in a way that drives culture on.

Such people also tend to be the ones with enough self-belief, enough passion about their artistic vision, to shrug off criticism and setbacks. Yes, it will be hard to see a Hollywood studio handing them $200 million to realise their dreams – but the avenues for creative self-expression fostered by the great acceleration, the zero-cost tools and distribution systems, are hardly going to disappear.

And whoever is producing them, this brave new world will by definition deliver entertainment products that we find even more compelling and appealing than today's: whether or not they are made by a computer or a human, they will be perfectly tailored to suit our tastes.

The paradox of acceleration, as we have seen throughout this chapter, is that while it renders culture speedier and more superficial, it also provides space for complexity and quality to shine through. It also produces more people who are clamouring to be part of the cultural conversation – and gives them more chances to do it.

In 1899, the painting *His Master's Voice* depicted new technology rendered comforting and familiar – in the form of a dog cocking an ear at a miraculous new device called a gramophone. So appealing was the image that it became the logo for a host of associated companies: RCA in America, EMI and HMV in the UK, JVC in Japan.

More recently, a rather different company has attempted to create its own idealised image of media consumption. The logo for Amazon's Kindle eBook reader shows a silhouetted figure sitting against a lone tree, lost in contemplation of the literary riches before him. You cannot even tell that what he is holding is a Kindle itself, rather than a book: the point is to erase the distinction between very new device and very old pleasure. Even in a world of speed and shocks, it seems to be suggesting, Philip Roth's cultural legacy is safe for a while yet.

5

Tomorrow's News Today

'Clearly journalism has got faster. Has it got better? Bearing in mind that the answer is "No".'

– John Oliver to Tom Brokaw, *The Daily Show*[1]

'Humiliated and hooded, the tyrant faces his fate on the steel scaffold.' This was the front-page headline with which the *Daily Telegraph* announced the death of Saddam Hussein in December 2006. The piece described how American and Iraqi officials had watched as Saddam met his end: 'The clang of the metal door was accompanied by a clunk, as the weight of the tyrant's body pulled the rope tight.'

There was just one problem: it wasn't true. When news of Saddam's death broke at approximately 3 a.m. UK time, a preview piece based on a briefing from Western officials about what could be expected was jury-rigged into what read like an eyewitness report.[2]

But by the time the article had been edited, typeset, sent to the printing presses, then delivered to Britain's breakfast tables, it was clear that something very different had taken place. Shaky footage leaked onto the internet showed the true execution scene, in which a jeering mob of leather-jacketed Shia thugs chanted slogans as Saddam was sent humiliated to his death. There was not a hint of the expected sombre atmosphere, or of the presence of officials, or of the hood mentioned in the *Telegraph*'s headline.

In the news industry, there are two commandments: to get the story right and to get the story fast. Ever since Samuel Morse sent his second message by telegram ('Have you any news?'), or Julius Reuter strapped his first market report to a carrier pigeon, reporters of every stripe have lived or died by being first with the facts.[3]

Yet as the great acceleration gathers pace, there is a growing feeling that the world is moving too quickly for any of us to keep up – even those whose job it is to report and explain the world's events. What was emblematic about the *Telegraph* story, for example, was not that the paper (where I then worked) got its facts wrong: every news outlet on the planet has made similar lapses, even if they are inevitably a cause for bitter regret. It was the speed with which the truth emerged, as both the *Telegraph* and the entire Western media were scooped by an Iraqi with a cellphone.

It is not just the nature of reporting that is changing. As audiences move online, traditional business models and ways of working are collapsing. Both the production and consumption of news often seem to be governed purely by speed, with old-fashioned 'stories' replaced by 'content' designed not to inform, but to feed the public's insatiable demand for digital diversion. Overwhelmed by the flood of information, both journalists and audiences are losing their ability to deliberate and discriminate.

Yet there is a parallel and equally powerful case that the acceleration of the news industry, while chaotic and often cataclysmic for news providers, has resulted in an explosion in quality journalism. As with popular culture, there are engrossing long reads and important investigations being produced alongside the quizzes and clickbait – arguably in greater profusion than ever. This chapter will not only examine the convulsive changes affecting journalism – the profession in which I have spent my entire career – but explain why I still feel so optimistic about its future.

SQUEEZING THE PRESS

There are certainly plenty of voices arguing that the current model of breaking news is, in a word, broken. Old-school newspapermen publish despairing books with titles such as *Blur*, *No Time to Think* or *Flat Earth News*, lamenting the death of proper reporting standards.[4] The verdict from 1999 of media critics Bill Kovach and Tom Rosenstiel still holds true: the news business 'is in a state of disorientation brought on by rapid technological change, declining market share, and growing pressure to operate with economic efficiency'.[5]

But these pressures have not sprung from nowhere: they are inextricably driven by the trends we have already been exploring. Readers have shifted away from buying newspapers because it is far more convenient to get the news delivered instantly via their computer or, increasingly, their phone. But they have also done so because there are far more calls on their time – both the array of cultural delights described in the previous chapter and the increasing demands of their working, social and family lives.

Ben Bradlee, the great *Washington Post* editor, is best known for overseeing the Watergate investigation. But he was just as proud of another accomplishment: creating the paper's Style section.[6] This became a bedrock of its circulation – and profits – because it gave every Maryland housewife the feeling that she had a connection to the DC glamour set. Yet recently, these loyal female readers, aged between 18 and 49, started disappearing. A working group, including Brigid Schulte, was formed to find out why. It turned out it was nothing to do with the quality of the paper – they just didn't have time to read it any more.[7]

The next problem is that the basic business model of newspapers has also been disrupted. What they traditionally offered was a bundle of information – news, theatre reviews, opinion pieces, TV listings, obituaries, classified advertising, puzzles, weather forecasts. But online, there are people out there who are doing each segment better than generalists can. If you're a TV nerd, do you read the *Independent*'s reviews, or Digital Spy's? If you've got a sofa to get rid of, do you place a small ad or sling it on eBay?

The result of this is that both print readership and revenue have collapsed. In America, newspapers' advertising income is now less than half of what it was a decade ago.[8] Inevitably, there have been cuts: the number of newsroom staff in the US, according to the American Society of News Editors, is down to 32,900, compared with 56,900 in 1990. In 2014–15, the total fell by more than 10 per cent.[9] The problem is not readership per se: many papers have developed online followings far in excess of their print audience. But the revenues from such an audience, whether via subscription or display advertising, have yet to match those of the traditional model – and may never do so, especially given the difficulty in selling ads on mobile phones, which is where the readers increasingly are.

For those in the media, this only sharpens the sense of turmoil and dislocation produced by the great acceleration: after all, very few of us can be certain that our jobs, or even our companies, will exist in a decade's time. The paradox, however, is that the same forces which have dried up newspapers' traditional revenue streams have massively increased the demands on their staff.

The evidence, says Nic Newman of the Reuters Institute for the Study of Journalism, is that 'audiences increasingly want news on any device, in any format, and at any time of day'.[10] This, in turn, has accelerated the news cycle. It is no longer enough for journalists to gather information and write a story for the next day's paper: instead, stories develop in real time, with the initial release of the information (usually on Twitter) being followed by a bombardment of comments, criticism or facts that shed new light on the story.

This can be feverishly exciting, but the need to publish to different streams and channels – to tweet the story out then put up a quick breaking news post for the website, then maybe do a blog, then take in the Twitter reaction, then finally produce a longer version for the next day's paper – represents a significant burden. In most newsrooms, a shrinking number of journalists are being asked to produce an ever greater volume of copy, with ever shorter deadlines: a Cardiff University study estimates that the workload of the average Fleet Street journalist has tripled since 1985.[11]

There have been various consequences of this shift, all of them profound. For one thing, if you ask fewer and fewer people to do more and more work, the inevitable consequence is that a slapdash quality may creep in. 'The pace has gotten dizzying for me and my colleagues just in the past few years,' claimed Howard Kurtz of the *Washington Post* a couple of years ago. 'Everybody wants it now-now-now. And that's understandable in a wired world. But the sacrifice is clearly in the extra phone calls and the chance to briefly reflect on the story that you're slapping together.'[12] Many journalists will sympathise with the wry comment of Barry Bearak, a *New York Times* reporter faced with a prison sentence for reporting illegally from Robert Mugabe's Zimbabwe: 'Really, anything is better than having to file four stories a day for the website.'[13]

This faster online news cycle, with its endless stream of reports, also helps to create a palpable sense that something is always happening, and that it must be responded to and reported on. John F. Harris, co-founder of the Washington news site Politico, observes:

> every technological and business innovation in media has tended to speed up the news cycle, and accelerate readers' and viewers' demand. In the last several years, that acceleration has reached exponential level, so that even what we faced six years ago, when we launched Politico – the demands are of a different order of magnitude now than they were then.[14]

The competition for traffic between the digital giants has spurred this process on. For example, it used to be the case when I was at the *Telegraph* that the Google News algorithms rewarded you for taking the time to put together a concise summary of a breaking news event: the highest traffic would go to 200-word stories that contained all the essential information.

But then Google became worried that its news service was not instant enough – especially compared with Twitter. So it switched its algorithms so that it was better for news companies to put out a single-sentence version of a breaking story as quickly as possible, then flesh it out later. Our newsroom duly followed suit.

With this pressure to publish, it is all too easy to take others' reports as fact – or not to bother checking, because there are few incentives to do so, or punishments for not doing so. There is also, as the *Guardian* journalist Nick Davies has trenchantly argued, far more of a temptation to simply regurgitate the messages provided by PRs, whose numbers relative to journalists have more than tripled since 1980.[15]

This is a particular problem when it comes to online news, whose very instantaneity sometimes leads to a sense – both among readers and editors – that it does not particularly 'matter'. There is now an entire industry based around scouring local papers in all parts of the world, or the far corners of social media, for weird, wacky and wonderful stories that will rack up shares on Facebook, once given an appropriately salacious headline. Whether they centre around a spurious survey, a piece of bogus science, a ghostly apparition or simply a feuding husband and wife, such stories – if sufficiently clickable – appear in their dozens on the websites and social media feeds of some of the planet's largest media companies. Nobody, at any point along the chain, is incentivised to actually check that these tabloid tales are accurate: why ruin the fun?

In an internet economy in which page views are the currency, getting something – anything – up online can often be the dominant imperative, even if it means covering or hyping up stories that aren't really stories at all. These are what people generally refer to as 'clickbait' – stories primarily designed to 'go viral' by being shared on social media rather than to convey actual information. A classic example is the case of the 'balloon boy', a child from Colorado said by his father to have become trapped in a helium balloon, which drifted to a height of several thousand feet: the story was a hoax, but still provided an entertaining few hours of coverage for the news networks.[17]

Such stories are popular not just because they are sensational, but because they are easy to cover: you only need to repeat what everyone else is already reporting. That lowers the barriers to entry, allowing sites such as ViralNova to amass audiences in the tens of millions with only a handful of staff by picking the right stories and crafting the right, ultra-shareable headlines.[18] This, in turn, drives many traditional media organisations to feel they have to compete on the same terms.

Take the tale of the Spanish zookeeper dressed in a gorilla suit as part of a training exercise, who was accidentally shot with a tranquilliser dart. The story went round the wired world, delivering juicy traffic to a host of blue-chip news organisations.[19] Yet none of them had actually called the zoo or the hospital mentioned in the reports: when one of my more experienced *Telegraph* colleagues did, he was told that yes, someone had accidentally discharged a tranquilliser rifle and hit a colleague, but the gorilla suit was a pure fabrication.

Such an ecosystem is obviously vulnerable to manipulation. In his book *Trust Me, I'm Lying*, the self-confessed 'media manipulator' Ryan Holiday (former marketing director for the controversial US firm American Apparel) describes how easy it can be to feed this beast by placing stories with small blogs that percolate upwards through the media ecosystem, with no one really bothering to check their bona fides because they don't have the time.[20]

In one memorable stunt, he uses the PR site 'HelpAReporterOut', which links self-proclaimed experts with journalists writing stories with predetermined angles, to test the limits of the possible.

> In fewer than four months, I was a long-suffering insomniac on ABC News, a victim of germ warfare on MSNBC, a witness to an inter-office romance on CBS, a reluctant stock market investor on Reuters, and, the coup de grace, a vinyl record enthusiast on the *New York Times*. I've never owned a vinyl record. In fact, I didn't learn what 'LP' stood for until I read about it in the very article I was quoted in.[21]

I myself played a similar role in keeping the viral news machine whirring, if only accidentally. One morning in July 2015, my wife was browsing Mumsnet, the parenting website. She discovered a small thread devoted to the question of whether Jeremy Corbyn, then the insurgent favourite to win the Labour Party leadership, was sexy.[22] Most people thought not, but there were a few funny answers – one said he was 'attractive in a world-weary old sea dog sort of way', another that 'If you half fancied Dumbledore, Corbyn is probably in the same area'. So I jokingly suggested to an ex-colleague at BuzzFeed via Twitter that this could be her next piece, since she'd been the first to uncover the 'Milifandom' (the band of teenage girls who had developed swooning, semi-ironic crushes on Labour leader Ed Miliband during the 2015 election campaign). BuzzFeed then tweeted out a screenshot from its UK politics account, followed by a pun about 'Jeremy Phwoarbyn'. By lunchtime, there were half a dozen articles on big news sites reporting on the Mumsnet thread, and Corbyn himself was being asked for his verdict. His status as a sex symbol had officially become 'a thing'.[23]

SPEEDING UP, DUMBING DOWN

For some critics, this remorseless churn of stories and semi-stories represents the final fulfilment of Joseph Pulitzer's prophecy that 'a cynical, mercenary and demagogic press will produce in time a people as base as itself'.[24] Instead of getting HD news, goes the argument, we get ADHD news – 'dumbed down and tarted up', as the US newscaster Dan Rather puts it.[25]

It is certainly true that there is an awful lot of crap out there – and that the newsgathering process is in painful transition between old models that can no longer be sustained, and new ones that have yet to be proved.

But as the examples above help to show, the heart of the problem is not that traditional journalism is being left behind by the online world. Rather, it is that old media, new media and social media often feed each other's worst instincts.

Nowhere was this more glaringly apparent than in the wake of the Boston bombings of 15 April 2013. In an effort to be first with the latest developments, a host of news organisations went far beyond what was sensible or reasonable. CNN, the television channel that originally launched the rolling news revolution, scaled new heights of breathlessness in its determination to provide second-by-second updates on the manhunt for the perpetrators. As the police tracked the bombers to the suburb of Watertown, and a gunfight broke out, the channel divided its screen into four, each with a correspondent standing on a different street, none of them knowing what was going on. Without a studio anchorman to tether proceedings, each babbled mindlessly over the others, waving their arms desperately to get the producers' attention. As John Oliver put it on *The Daily Show*, 'they were literally talking faster than they were thinking – they had removed their internal monologue and were just throwing words at the camera'.[26]

Of course, in the immediate aftermath of any traumatic event, there will always be a scramble for detail, in which rumours spread via Chinese whisper before gradually being confirmed or rejected. For example, the *Wall Street Journal*, which generally reported on the bombings with clarity and restraint, briefly suggested that there had been five, rather than two, explosive devices planted at the marathon's finish line.[27]

Some, however, made their mistakes with unapologetic elan – not least the *New York Post*. Soon after the explosions, the paper reported that 12 people had died, a line it stuck to even as the rest of the media corrected course to report the real figure of three. Next, it exclusively revealed that a 'Saudi national' who suffered shrapnel wounds at the scene had been taken into custody at a nearby hospital by police: it turned out that while the man had been questioned, he had been cleared as a suspect.

Then came the greatest blunder: the decision to print, on the front page of 18 April, a picture of two men wearing backpacks whose images were being circulated by law-enforcement officials (along with many, many other such pictures) with the headline: 'Bag Men – Feds Seek These Two Pictured at Boston Marathon'. A weaselly caption noted: 'There is no direct evidence linking them to the crime, but authorities want to identify them.' But the implication was clear: here were two brown-skinned men who might very well be the perpetrators.[28]

Sadly for the newspaper, the duo – running enthusiasts named Salaheddin Barhoum and Yassine Zaimi, one of them just 17, and therefore a minor – had already made contact with the police, after images of them began to circulate online. They had also (according to the lawsuit they subsequently filed against the *Post*) been told they were not suspects. Their lawyers alleged that the paper had, in effect, been willing to ruin their lives in pursuit of a cheap scoop – to put a great news line above any ethical duty.[29]

The *Post* was, of course, an outlier: I cannot imagine that front page being approved, or even seriously suggested, in any newsroom I have ever worked in. Yet its publication may not have been the result of pure sensationalism. Increasingly, as news migrates online, papers feel they have to shout louder, push harder, in order not just to drive the story forward, but to remain relevant.

Indeed, many people on the web treated the Boston bombings not as a news story, but as a test case of whether the speed of online culture had finally rendered old media completely irrelevant. 'If Sunil Tripathi did indeed commit this #BostonBombing, Reddit has scored a significant, game-changing victory,' tweeted a hitherto anonymous Twitter user called Greg Hughes at one point. Later, he added: 'Journalism students take note: tonight, the best reporting was crowdsourced, digital and done by bystanders. #Watertown.'[30]

What Hughes was referring to was a colossal online effort, taking in Twitter and a thread on the popular website Reddit called 'find-bostonbombers', which sought to bypass the clumsy, slow-paced efforts of traditional journalists – and indeed law enforcers. Instead, it would track down the bombers by harnessing the frenetic energy of the online group mind.

The impulse – similar to the response that followed disasters such as the Haiti earthquake – was entirely admirable, as was the list of guidelines that the creator of the Reddit thread, an English poker pro called 'oops777', set out.[31] This was not to be a vigilante effort; any personal information was to be shared with the FBI rather than posted on Reddit; racism and speculation would not be allowed; everyone would be innocent until proven guilty.

These rules were sensible and well-meaning, but all were comprehensively flouted. Instead of a Sherlock Holmes-style feast of deduction, what developed was essentially an online lynch mob. On both Reddit and another site, 4chan, suspects' names, details and ethnicities were thrown about pell-mell, in what one commentator described as a racist game of Where's Wally?[32] Inevitably, these names spread beyond the discussion forums: either taken up by members of the mainstream media desperate for a lead, or by those determined to subject those mentioned to impromptu campaigns of online vilification.

One name that cropped up frequently was Tripathi's. He was a student at Brown University who had gone missing in March, leaving behind a cryptic note, and someone who went to high school with him thought she had identified him in one of the pictures from the bombing.

This is where Hughes comes in. A part-time journalist and software engineer, who has since erased his Twitter account (and much of his online presence), he was up in the early hours of 19 April, following the latest developments. It was a dramatic time: police had cordoned off large stretches of Watertown, and were engaged in a running gun battle with the bombers.

According to the timeline reconstructed by the US journalist Alexis Madrigal, at 2.14 a.m. an official using the Boston police scanner read out an ID, which ended: 'Last name, Mulugeta, M-U-L-U-G-E-T-A, M as in Mike, Mulugeta.'[33] It was noted by the online observers. At 2.42 a.m. Hughes tweeted: 'This is the Internet's test of "be right, not first" with the reporting. So far, people are doing a great job. #Watertown.'

Then, a minute later, Hughes tweeted the following: 'BPD has identified the names: Suspect 1: Mike Mulugeta. Suspect 2: Sunil Tripathi.' This was, as Madrigal pointed out, three errors for the price of one. The original scanner request had said M 'as in Mike' – not that the suspect was named Mike. Nor, for that matter, had it said this Mulugeta was a suspect. And Tripathi's name had not even been mentioned, although he had figured heavily in the unofficial speculation that had been taking place online.

It didn't matter: this lie went round the world before the truth had got its boots on. First tens, then hundreds, then thousands of tweets spread the news. Many rejoiced that the Redditors' efforts had been vindicated and that 'old media' had been outflanked. Among them were the 'digilantes' who decided to vandalise a Facebook group set up after Tripathi went missing; his family received obscene and pitiless threats.

But it was all a grotesque mistake. The real Boston bombers were the Tsarnaev brothers, one of whom – Tamerlan – had already died of injuries sustained at the hands of police marksmen before Mulugeta's name even came over the scanner. Perhaps the most horrible moment came when a body, identified as Tripathi's, was found in a nearby river shortly afterwards. For a brief moment, it seemed as though the online hate campaign had driven him over the edge, until it was confirmed that his suicide had taken place well before the bombings.

In this episode we see summarised many of the changes in how we consume – and produce – our news, in which our craving for speed defeats our better judgement. There is also the fact that the range of voices that can make themselves heard has expanded exponentially – a good thing in that it widens the debate beyond the traditional oracles of newsgathering, but a bad one in that it opens the door to misinformation too (as well as outright propaganda: see the al-Shabaab terrorists who took control of a shopping mall in Nairobi in September 2013 and used a series of Twitter accounts to update the world and taunt the police, or the many accounts tweeting in support of or on behalf of Isis).[34]

Boston was merely one example: in the wake of any major incident, it is all too easy to give currency to false information. In the case of the mass school shooting in Newtown, Connecticut, on 14 December 2012, the idea was circulated that the guilty party was the (entirely innocent) brother of the actual shooter, Adam Lanza.[35] Often, an echo-chamber effect takes hold, as the amateur reporters of the web feed on and amplify the media's mistakes – or fill the gaps that the mainstream media cannot, due to legal constraints. It was the BBC's *Newsnight* that falsely accused Lord McAlpine, the former treasurer of the Conservative Party, of paedophilia, on 2 November 2012. But it was the rumour-mongers of Twitter – including the wife of the Speaker of the House of Commons – who actually named him after a loose-lipped BBC executive tweeted the news of the impending 'scoop'.[36] Social media was also used to demonise the teenage suspects in a rape case in the town of Steubenville, Ohio in early 2013, even as the authorities were powerless to comment on an ongoing investigation and court case.[37] (On a less grave note, Twitter also helped a British comedy writer, Graham Linehan, to persuade large parts of the world that Osama bin Laden liked to watch his sitcom, *The IT Crowd*, in his lair in Abbottabad.)[38]

TALKING HEADS

So far, this catalogue of media calamities has focused mostly on the written word, whether in print or online. But as the Boston bombings saga suggests, there is an equally severe problem when it comes to news on TV. Indeed, before the internet even existed, TV was driving forward the process of acceleration. When CNN launched, say Bill Kovach and Tom Rosenstiel, 'every new interview or factoid seemed to be a scoop. Every few minutes presented a new deadline. It was an adrenalin rush. The news seemed organic.'[39] The network pioneered techniques such as the 'Breaking News' tag, and the 'crawl' of constant updates on the bottom of the screen, which seemed to deliver a new dimension of instantaneity.

Yet now we have moved from the 24-hour news cycle to what *The Onion* has lampooned as the '24-second' news cycle.[40] In such circumstances, the then president of CNN US told authors Howard Rosenberg and Charles Feldman, 'crawls are almost antiquated. They seem almost sluggish in comparison to how quickly a website loads or changes pages . . . Our bodies are so revved up now that the crawl itself feels almost stately . . . and we try to acknowledge that by cutting our cameras more quickly, putting up more images.'[41]

In terms of delivering information, this process is hugely counterproductive: as we saw with multitasking, the human brain is incapable of concentrating on so many things at once. When, in a fascinating study, academics stripped away much of the graphical gimmickry from rolling news, information retention shot up.[42] But that is to ignore the obvious: that the purpose is not to inform, but to keep you watching, and ideally to do so at the lowest possible cost per eyeball.

This is the traffic-driven model of newspapers, applied to the screen. Indeed, the shrinkage in ambition has, if anything, been even worse. On American local news, a combination of shorter attention spans and lower budgets has seen story length collapse, while sports, weather and traffic have expanded to fill 40 per cent of the schedules.[43] At CNN, the length of time devoted to properly filmed and edited news reports halved between 2007 and 2012, replaced by less expensive but less informative in-studio discussions.[44] Like print journalists, those reporters that remain have seen their workload increase: instead of taking the time to polish pieces, they now have to file to the morning, lunchtime and evening news, perhaps with a few radio broadcasts or blogs for the website thrown in. Instead of being free to rove and report, they are the prisoners of their cameras and satellite phones – resulting in those laughable scenes in which a reporter standing for pictorial purposes outside a courtroom or government department that has long since been shuttered for the night has to be informed by the anchor of the latest developments.

In place of reporting, the networks have turned to opinion. To paraphrase the great *Guardian* editor C. P. Scott, facts are expensive, but comment is cheap: you don't need a colossally expensive outside broadcast unit or a pricey foreign bureau to stage a shouting match between two commentators, or have your anchor deliver his own immaculately coiffed verdict. On America's worst offender, MSNBC, such programming made up 85 per cent of content in 2013.[45]

Of course, opinion journalism is not necessarily a bad thing (which is what you'd expect me to say, as a former comment editor) – so long as you actually present the facts as well as scrapping over them. The trouble is that on live TV, it is far harder to challenge guests, or muster contrasting evidence, not least when the presenter is flitting between topics with little time to prepare. With most discussions hovering around the three-minute mark, Kovach and Rosenstiel argue that 'most live interviews are less a revelatory encounter than a kind of ceremonial ritual'.[46] The bite-sized format also means it is harder to get complex or nuanced thoughts across.

It doesn't help matters that the key factor that marks a commentator out for stardom is not genuine expertise, but plausibility (or, failing that, aggression). Certainly, these pundits aren't being hired for any genuine prophetic expertise: in a landmark study, the American sociologist Philip Tetlock evaluated thousands of predictions by nearly 300 political experts over the 20 years from 1985 to 2005, and concluded that most of them would have been bettered by a 'dart-throwing chimpanzee'.[47]

The increasing prevalence of what might be called second-order journalism – people talking about the news rather than just reporting it – has also given rise to another form of acceleration, across all forms of media. To the everlasting regret of editors everywhere, there are natural constraints on how fast news can go – how long it takes to send a reporter or cameraman to the scene, or for someone to respond to a rival politician's statement or accusation in a way that puts a fresh spin on the story.

But so great is the public's, and the media's, addiction to speed that we have come up with an ingenious way round such limitations: by delivering news not from the present, but from the future. No longer is it the done thing to wait until someone actually says something before you report it. On both sides of the Atlantic, political speeches are now briefed well in advance, with coverage shifting seamlessly from the political implications of what will be said to the backlash against what has been said with barely a pause to report what was actually said. As John F. Harris says, 'By the time the President actually utters something, we've chewed it over in anticipation for so long that we've already moved on.'[48]

This phenomenon has an especially interesting consequence for newspapers. If a story breaks in the morning, by the time it's in the next day's edition it's already almost 24 hours old. Will people really buy a front page that tells them what they could have learned on the evening news? Not in the opinion of one senior member of the Downing Street team. He believes that 'the fact that the story of the day has churned through so many iterations during the day puts more pressure on newspapers to be sensationalist, to have a new angle on each story. The newspaper as reportage, as straightforward reporting – no one's really interested in that any more. It's created a kind of hecticness.'[49]

There is certainly a strong argument that the great acceleration has created a coarse, chaotic culture, in which a slicker, shallower and ever more speed-obsessed media generates much heat but little light. And matching the pattern seen elsewhere, there has also been a shift towards gigantism. To capture the viewer or reader, news either has to matter to them personally, or to everyone. Indeed, in an age of splintered audiences, news organisations long for big stories, epic events which can sell papers or attract viewers by piercing the background hum of day-to-day trivia.

This is why CNN, for example, offered carpet-bomb coverage of the Oscar Pistorius trial, or the hunt for Flight MH370, the Malaysia Airlines plane that disappeared mid-flight: once the statistics told editors that viewers were interested, they were fed all they could stomach and more. Even for smaller stories, it is easier than ever to give the readers what they want. Whereas a print newspaper can clear a double-page spread for a big story – but would still have to fit the rest of the world's news in – an online publication can see traffic spiking up for a particular story (say, an American dentist having killed a lion called Cecil in Zimbabwe) and commission an avalanche of further pieces on the topic, in the hope that one of them will take off. In February 2015, the world went briefly nuts over a dress at a wedding which looked both white and gold and blue and black. The original BuzzFeed story – a viral phenomenon by any standard – received approximately 40 million hits. But the follow-up stories which the site's US and UK teams threw themselves into producing garnered tens of millions more.[50]

THE CASE FOR DISRUPTION

Given the apocalyptic picture painted above, it may seem surprising that I and others not only remain in news journalism, but remain convinced that we are living in the most exciting, ambitious and transformational period in our profession's history.

The truth is that the speed of the news agenda, and of the disruption of business models, has brought huge challenges to journalism. Every day throws up new questions, not least over the relationship between the media and the internet companies. Recently, there has been much fretting over the power handed to the giant gatekeepers such as Facebook – to the point where Greg Marra, the twentysomething who curates its News Feed, is probably the most powerful single individual in the journalism industry.[51] If Facebook's algorithms decide (as they did) that race riots in Ferguson, Missouri, are less interesting than celebrities taking the 'Ice Bucket Challenge', then people will see the ice but miss the riots.[52]

Again, aspects of the great acceleration are feeding on themselves. Facebook needs to cater to people's demands for instant gratification. That means cutting the appallingly long loading times on portable devices – the average mobile web page, cluttered with advertising and hidden software, takes eight seconds to load, aka an eternity.[53] The solution is to host stories within its own app – via its 'Instant Articles' and 'Instant Video' features – so that there is no loading time. But this gives Facebook's platform enormous power over the publishing industry, to the point where some fear it will capture all the value and leave news companies as commoditised, subservient appendages.[54]

Already, Facebook is changing the way we consume news simply by existing. Network effects mean that the stories that are shared most do an order of magnitude better in traffic terms than the second-most-popular pieces – so the economic imperative, as with publishing or music, is for news companies to focus their attention on pieces that will go viral and neglect the bread-and-butter news stories. It also means focusing on happy, uplifting stories, since these are more likely to be shared – in sharp contrast to the old model of focusing on stories designed to terrify or enrage your readership.

Then there is the question of whether Facebook or Twitter outrage or interest actually translates into anything meaningful. When the Islamist group Boko Haram kidnapped hundreds of Nigerian schoolchildren, the world embraced the hashtag '#BringBackOurGirls' – but the girls stayed kidnapped.

For editors and journalists, the online ground keeps changing beneath our feet. One of the wisest observations comes from the *New York Times*'s landmark strategy review, which refused to talk about a 'transformation', because 'it suggests a shift from one solid state to another; it implies there is an end point'.[55] Even as the *NYT* mastered the web, readers were switching to phones and tablets; even as it was redesigning its home page, traffic was shifting to specific stories reached via social media.

The crucial thing about this scary new world, however, is that – while there is a tremendous amount of dross – there is also far more good journalism being produced than at any time in history. For example, newspapers have spent the past decade or more in the throes of digital convulsion, racing against time (and declining circulation) to transform their newsrooms into temples of speed, driven by the throb of Twitter and lit up by projected charts giving instantaneous feedback on how particular stories are performing.

Yet even as they have adapted to cater to the voracious demands of the online audience, newspapers in Britain and elsewhere have also produced a string of scoops – many largely enabled by technological change and data analysis – that arguably eclipse those of any other era. The *Telegraph* famously obtained the expenses details of Britain's MPs, prompting a national political scandal – but it has a host of other investigative achievements to its name, ranging from the exposure of regulatory failings in the NHS to exam boards prepared to help teachers game the system. Then there were the *Guardian*'s revelations, via Edward Snowden, about the National Security Agency's programme of online surveillance. The *Washington Post*, *Wall Street Journal* and *New York Times* are also competing as fiercely as ever for their Pulitzer Prizes, and making those in power just as uneasy. And new players, such as my former colleagues at BuzzFeed, are ploughing significant resources into investigations.

In short, if one of the main functions of journalism is to find out and publish information that powerful people would rather keep hidden, no one can argue that the press is not doing its job. Back in 2008, professional journalists – a frequently curmudgeonly bunch – told researchers by a margin of two to one that straddling print and online has made them better at their jobs: better able to find out facts, to respond to criticism, to engage with their audience.[56]

With information being generated, and moving from place to place, more and more rapidly, readers can now get more information on virtually any subject than at any time in history – and better information, too. Indeed, as Tom Standage of the *Economist* has written, 'although the internet has proved hugely disruptive to journalists, for consumers – who now have a wider choice than ever of news sources and ways of accessing them – it has proved an almost unqualified blessing'.[57]

Those who condemn the new world of instant reaction and comment are often confirming their own pessimistic view of human nature. Yes, it can be disturbingly easy to whip up a mob or spread a falsehood. But then, it always has been: one of Chaucer's earliest poems, *The House of Fame*, is largely concerned with the dangerous power of malicious gossip.[58] Today, it is also easier to speak truth to power: to mobilise boycotts of companies that have dodged tax, or highlight errors in the mainstream press. The web makes what Nick Davies calls 'churnalism' more tempting – but it also helps people construct automated software that detects those who simply recycle others' words.

Similarly, one of the oldest principles in journalism is that there's always someone closer to the story than you – but with the internet, it's not only easier to track them down, but for them to make their voices heard without your help. CBS News, and its anchorman Dan Rather, claimed that George W. Bush had been given preferential treatment during his National Guard service, which helped him avoid serving in Vietnam. The story was exposed as a fake partly due to experts on 1970s typewriters pointing out errors in the font and spacing on the documents on which the story rested.[59] Such outside experts can not only critique the news, but generate it: Eliot Higgins, a blogger from Leicester, was the first to confirm that Syria's president Bashar al-Assad was using chemical weapons on his own people, after making himself a world-renowned expert on the regime's weapon systems and a distributor for online videos smuggled out by the opposition.[60]

Even the great debate over how to make news pay (or, at least, pay as much as it used to) has obscured Standage's point that the essential problem is excess of supply – that there are too many things to read, not too few. Take a statistic from the 2013 Pew survey of the state of American journalism, often cited as a harbinger of doom: that 31 per cent of Americans have deserted a particular news outlet because it no longer provides a good enough service.[61] Does this mean that quality is falling? Perhaps. But it also means that the market is at work. Those people have not simply stopped consuming news – they are getting it from places that are more convenient, or more informative.

QUANTITY AND QUALITY

For traditional news organisations, the blessing of the internet is that you can peddle your wares to everyone in the world. But by the same token, you are suddenly in competition with the world, too – not just with rival newspapers or TV networks, but with a Google algorithm, or sites such as Reddit or Slashdot whose content (and the ranking of that content) is generated by their users. You are also, as publishers and others have found, in competition with every other entertainment product in the world for people's attention.

As in the culture industry, this leads to two separate reactions – a tendency towards attention-grabbing sensationalism, but also a corresponding drive towards quality. In breaking news situations such as the Boston bombings, there is still a vast volume of cautious, properly sourced reporting, as newsrooms scramble to send their best people to the scene. As elsewhere, the middle is being hollowed out. But the result is a creative kind of destruction: the emergence of a few vast global news organisations with the resources to carry out proper, serious journalism as well as the fun stuff, and niche outfits which offer exceptional coverage of a few key areas.

Consider *Politico*, mentioned above. The site, and its affiliated newspaper, was explicitly a response to an acceleration of politics that traditional media organisations seemed unprepared to deal with. Political coverage in the mid-2000s was, John F. Harris recalls:

> still dominated by a handful of big news organisations, which ran on very traditional rhythms. Over at the *Washington Post* [his then employer], of course there was a website, of course people would pay rhetorical deference to the fact that we needed to respond to the news in real time, but in fact, the rhythms of daily life were totally dictated by the same once-a-day print schedule.[62]

Politico made its name by speeding up the delivery of political news, not least through its early-morning Playbook email. It certainly still does its best to be fast: its unofficial motto is 'Win the Dawn', echoing the 'Win the Day' slogan adopted by Bill Clinton's campaign.

Harris admits that 'it requires a certain discipline for journalists not to be buffeted by every momentary sensation'. But he also argues that Twitter 'has turned almost everyone into a blogger, and there's almost no way to win the competition on speed grounds alone'. So, he says, 'we pride ourselves on being really quick to respond to the news, on being a 24/7 newsroom, but it's not possible to stand out on the basis of speed – you've got to have a qualitative advantage in some way. The way to stand out is through deeper, more analytical enterprise.'

That is a sentiment shared by Ben Smith, a Politico veteran who made his name by running one of the blogs it produced to cover the 2008 election – in itself something of an innovation. The format succeeded because it met users' craving for speed: 'I'd go away for a couple of hours, and start getting emails asking whether I'd died,' says Smith.[63] But then the acceleration continued, and Twitter – with its near-instantaneous updates – made even blogs look cumbersome: Smith sensed the momentum, and conversation, moving elsewhere.

Yet like Harris, he came to realise that chasing the demon of speed down the rabbit hole wasn't the recipe for success. Instead, it was to combine agility with quality. Smith is now the editor-in-chief of BuzzFeed (and as such, my former boss). To its detractors, this is a place for mindless features with titles like '21 Flawless Cattacks' or '22 Celebrities You Probably Didn't Realise Were The Same Age'. But beneath the snarky 21st-century veneer, BuzzFeed is that most old-fashioned of creatures, an outlet for interesting journalism, in the form of snappily produced, often surprisingly serious pieces that practically beg you to read them. It does the instant, and the considered, with equal verve and quality. Rather than pursuing a particular niche, BuzzFeed is a mass-market proposition with a vast and growing international audience which relies on its style and voice as its USP – and, of course, the quality of its technology and writing staff.

Indeed, when discussing BuzzFeed's success, Smith makes the same point that was raised earlier about the way the internet works. 'The transparency of Twitter and the social web means that the best thing will do way better than the second-best thing,' he says. 'The stuff that succeeds is usually the best, and you have to hire the best people to do it.'[64]

It is easy to mock BuzzFeed's viral news posts if you have not seen the craft that goes into them – the determination to treat these as proper news stories, to get the extra quote or chase the original source so as to ensure that it is BuzzFeed's version which stands out from the crowd of online imitators. And added to this is a determination to think up new framings for stories, or tools to tell them better, in order to remain ahead of the chasing pack. It also has a native advertising model – making ads for its clients which resemble its stories and are shared in the same way – which appears, at least for the moment, to have overcome the long-standing problem of how to make online news pay.

What we are seeing is, in other words, exactly the same shift as in popular culture. Just as the single displaced the album as the atomic unit of pop music, so the story has replaced the newspaper. Increasingly, we no longer buy the *Telegraph* or the *New York Times* – we consume individual stories that reach us via our news feed, or any of a thousand other routes. That means that you have to focus on making every story as readable as it can be: in the words of MIT media professor Henry Jenkins, 'If it doesn't spread, it's dead.'[65] Often, that means putting a loud and hyperbolic headline on the story. But it also means – or can also mean – producing content so good it demands to be read.

Just as with books or films, the speed and power with which information is shared online does not benefit simply the hastiest and most hysterical stories (though it certainly does that), but the most compelling, too. Even BuzzFeed's infamous lists turn out to be evidence not of an ADHD culture, but of one in which people are willing to devote attention to those things that reward it. 'The lists that do really, really well for us,' says Smith, 'tend to be much, much longer than anything you'd find in a magazine.' It's not just lists. During my time there we published a fascinating 3,000-word story about someone who had stolen someone else's life on the internet, which received almost two million views.[66] A 6,200-word long read from the US team on buying a house in Detroit for $500 had more than a million readers – and held the attention of those using smartphones for an average of more than 25 minutes.[67] Overall, BuzzFeed pieces of 3,000 words or more are shared four times more than the average; the same holds true for other publishers such as the *Guardian*.[68]

What is disappearing, the evidence shows, is not the snappy, short update, the impeccably shareable gag, the provocative think-piece or the considered in-depth report, but all the stuff in the mediocre middle – the 600- or 1,000-word summaries of the day's events that exist to fill a newspaper's columns, rather than catch the reader's eye. And those great pieces, whether long or short, can now be shared across all manner of platforms more quickly than ever before.

Indeed, when Meeyoung Cha of the Max Planck Institute for Software Systems analysed 54 million Twitter accounts, he found that those who built the biggest audience were those with genuine expertise, who offered a regular stream of links and news on their pet topic.[69] As Clive Thompson put it in *Smarter Than You Think*, 'quality, delightfully, seems to matter'.[70] Similarly, research from the Reuters Institute shows that those it dubs 'News Lovers', who consume the most news and drive the agenda forward, are drawn to upmarket, intelligent reporting rather than celebrity gossip.[71]

In their book *The New Digital Age*, Eric Schmidt and Jared Cohen of Google argue that when everyone in the world has a smartphone, everyone will be a reporter: imagine that Iraqi at Saddam's execution, times a few billion.[72] Professional journalists will no longer be front-line information-gatherers, but information-sorters. And they might even be joined by celebrities and others who set themselves up as newsmakers, with their personal brands just as powerful, and just as trusted, as those of the old media stalwarts.

What we probably will have to say goodbye to in such a world is the media's role as referee of what we read and what we don't. Walter Lippman, one of the founding fathers of American journalism, observed that 'the news of the day as it reaches the newspaper office is an incredible medley of fact, propaganda, rumour, suspicion, clues, hopes and fears'.[73] The task of selecting and ordering that news, he went on, 'is one of the truly sacred and priestly offices in a democracy' – not to mention an editor's bread and butter. But now, that task has been superseded by the internet: what is your Twitter feed, after all, but a personally selected and ordered distillation of all the news you think is fit to print, delivered with an immediacy no paper or website can match? This filtering process may be messy – and occasionally, as with the Boston bombings saga, disastrous. But the collapse of the traditional media model under the pressures of acceleration has not been matched by a collapse in our ability to find out what is going on in the world. If anything, the reverse.

THE NEXT GENERATION OF NEWS

In the future, some journalists may be grungy hackers rather than shabby hacks, able to comb through vast databases as easily as their forebears combed through court reports or official statements. And the gatekeeping function may well transmute into community management, or simply expert curation: as data floods in from automated sources, from amateur reporters, commentators and conspiracy theorists, or from political campaigns or companies with a point to make or axe to grind, the editor's role will be to validate and select, to sift through in search of the nuggets on behalf of a time-poor public – or else to add value to that content via the injection of expertise, wit or opinion.[74]

Indeed, one benefit of the move away from TV and on to the internet is that it breaks down the monopoly of the news channels. In an earlier chapter, we saw that many of the problems we have with 'screens' in raising our children are, in fact, a problem with television – and the same is true in the media. The internet, in contrast to television, is a 'lean-forward' experience, which encourages you to be more active than passive. It is also an environment that privileges the human voice – people with lively, funny or interesting things to say – over stultifying corporate-speak, as well as breaking down barriers so that the talented have a far more powerful megaphone with which to make themselves heard.

The final thing to consider, in evaluating the increasing pace of the media, is what will happen as automation begins to affect not just the business structure of journalism, but its content. As in every other field, a host of entrepreneurs have appeared who are determined to bring disruptive innovation to the process of news-gathering. That may mean converting fact into story without the intervention of human beings, or else deploying algorithms rather than editors to determine what stories people want to read (and if the result is anything like what happened when Amazon put its staff of professional editor/curators up against its book recommendation algorithms, people like me had best start polishing our résumés).

In terms of machine-generated journalism, there are companies such as Narrative Science, a start-up based in Chicago that recently signed a deal with the CIA. It promises to turn big data into clear-English narratives, stripping out the pesky human element from journalism.[75] Similarly, Automated Insights of North Carolina produces thousands of corporate earnings reports for the Associated Press and fantasy football reports for Yahoo!, all written by robots rather than people. AP reckons this enables it to cover ten times as many stories as previously.[76]

Journalists, too, are adopting such techniques. When on 1 February 2013 a small earthquake – magnitude 3.2 – struck California in the early hours, the *Los Angeles Times* had a report by Ken Schwencke up within eight minutes. In fact, Schwencke had written the code for an automated script that detected an earthquake alert on the US Geological Survey website, and plugged the information into an auto-generated story.[77] It might seem like this process threatens jobs, but if basic stories – such as earnings or weather reports, or even simple accounts of sports games – can be computer-generated, it leaves journalists free to carry out more productive and higher-value tasks.

The same is true of Silicon Valley's attempt to automate the summarising and aggregation of news. Google News is already, arguably, an example of this. Then there is Summly, founded by a British teenager and bought by Yahoo! for roughly $30 million; this app promised to mine stories for the vital data, stripping them down to shorter and more easily digestible chunks on your mobile device.[78] Or Prismatic, one of many attempts at a 'serendipity engine', which gives you not just the most obvious stories, but offbeat and oddball stories that its algorithms suggest you might be interested in, whether they are published by the *New York Times* or a random student with a Tumblr account.[79]

Some of the sites and services that result may not look much like journalism. Bradford Cross, the founder of Prismatic, is a data scientist and former hedge-funder who built a website called FlightCaster, which predicts flight delays before airlines announce them. His co-founder at Prismatic was a PhD student in machine learning; two of their initial hires, as the US journalist Will Oremus pointed out on *Slate*, have PhDs in natural-language processing and artificial intelligence.[80] None of them, as far as it is possible to tell, has ever held a red pen, or even written a headline. But then, few editors have written software capable of crawling the entire English-language internet to analyse every story, as Prismatic has. And if the effect of their efforts is to give people better things to read, and spread good journalism to a wider audience, can we old-timers really object?

Well, some would argue – as they do of the Facebook News Feed – that the fragmentation of the media is causing society, too, to fragment, that all these helpful little algorithms will push us into our own little worlds, in which a 'filter bubble' shields you from seeing anything you might ever dislike or disagree with.[81]

Yes, if people never encounter contradictory opinions, instead clustering together in small groups that reinforce and affirm their beliefs, then society will indeed become polarised, as each group picks the fragments that it likes from the great flood of information. But the evidence suggests this is not happening, or at least that it is happening more slowly than the prophets of doom foretold.

Twitter, for example, is a personalised news feed, but it is also a diverse one, which exposes you to stories you would never have come across. As Steven Johnson points out in his book *Future Perfect*, a 2010 study by Matthew Gentzkow and Jesse Shapiro did find that online discussion had an echo chamber effect. But it also found that 'neighbourhoods, clubs, friends, work colleagues, family – all these groups proved to be deafening echo chambers compared with all forms of media'.[82] As personalisation becomes increasingly sophisticated, there will of course be a risk that the bubbles around us grow smaller. But the countervailing effect of the social web – the fact that everyone likes to read what everyone else is reading – will still work to pierce it.

There is, in the end, a frightening amount of information coming at us, at a frightening pace. Much of it is tacky, sensationalist or simply disposable (but then, there is a place in everyone's life for a few minutes of hilarious cat videos). Often, the media may – as with the Boston bombings – give in to its worst instincts, and there is little chance of that changing. But as we have also seen, the trivial and substantial, the instant and considered, can and will coexist.

As the world grows more and more connected, the volume of the information pouring in will only increase. For public and journalists alike, the world, and the news cycle, appear to be spinning more rapidly: terrorist attacks, economic shocks, moments of triumph and disaster come thicker and faster than ever. That will give us, more than ever, the feeling that we are trapped on a turbulent, scary planet, in which drama and crisis are never far away. For the old high priests of information, this process has been and will continue to be a traumatic one. But for the public, the ensuing torrent of news and diversion is one in which they are more likely to swim than drown.

6

The Pace of Politics

'The chief characteristic of the modern world is the scope and speed of change.'

– Tony Blair[1]

In Britain, the three candidates for prime minister are holding the first ever televised election debate. Below them, on the viewers' screens, crawls 'the worm' – a real-time index of how a focus group of demographically representative voters is reacting to their words. Say something particularly touching, and the line spikes up. Insult another candidate, and it ticks down.

In America, a campaign spokesman is being interviewed on the radio. On the laptop in front of him he can see the donation figures for his boss's website. He notices that they are going up as he speaks: he uses a particularly popular catchphrase and sees, after a few seconds, a spike in donations. In his inbox, meanwhile, sit the latest breakdowns from the campaign's IT people: they tell him that, from now on, all emails will conclude with a suggestion to 'Learn More' rather than an instruction to 'Sign Up', since in randomised testing, the former has outperformed the latter by almost 20 per cent.[2]

The effects of the great acceleration on the media have certainly been tumultuous. Yet it has had an equally powerful impact on the most frequent subject of the media's attention – the world of politics and government. It has provided opportunities for politicians to use technology to make more direct and responsive connections with the electorate, and for government to become more agile and more flexible. But mostly, it has subjected both the process of campaigning and the process of governing to a level of pressure that makes it ever harder to do either well. Indeed, we seem to be reaching a point of terminal dissatisfaction with the politicians – and the politics – that the great acceleration produces.

THE PERMANENT CAMPAIGN

Political campaigns these days are vicious, short-termist and superficial. Yet by historical standards, they are positively decorous. John Adams not only accused Thomas Jefferson of being a pagan, atheist and traitor, but his supporters claimed that, if Jefferson were elected in 1800, 'murder, robbery, rape, adultery and incest will be openly taught and practised'.[3]

What has changed is not what people are prepared to do to get elected, but that the ethos and tempo of electioneering are no longer confined to campaign season. Increasingly, politicians cling to them even when in the heart of government – and they do so because the only alternative, as they see it, is to be swamped by an ever more fickle and ferocious media. H. L. Mencken once claimed that politics is 'the art of running the circus from the monkey cage'.[4] Today, it can feel more like doing so from the piranha tank.

It was Bill Clinton's guru James Carville who insisted that politicians 'must always be ahead of the news cycle'.[5] But that commandment has now been raised to the status of holy writ. The late Philip Gould, Tony Blair's polling guru, described the art of modern election campaigning as one of 'dominating the news agenda . . . with stories and initiatives that ensure that subsequent news coverage is set on your terms'.[6] This means 'anticipating and pre-empting your opponent's likely manoeuvres, giving them no room to breathe, keeping them on the defensive. It means defining the political debate on your terms.'[7] It also means being willing to roll around in the gutter. As the Democrat strategist Joe Trippi points out, the political messages that do best at puncturing the filter bubble are *ad hominem* attacks. As a result, 'the most effective ads are the ones that make the community a worse place to live'.[8]

The result of all this is a style of communication driven by fear and aggression: fear that the story of the minute will blow up in your candidate's face, and aggression to ensure that the other side takes the brunt of the blast. 'The campaigns feel that they need to be fighting every moment of the day,' says John F. Harris, co-founder of *Politico*.

> They used to have a win-the-day mentality, which was largely dominated by who came out ahead on the evening news. Then they moved to a win-the-hour mentality, where emails and faxes were flying back and forth. Today's political campaigns feel they need to win the minute, with every charge responded to in real time. They view stories almost like forest fires – if you don't pay attention to the leaves and twigs that are burning right now, they can take out the whole forest in an hour. So they end up fighting back against stories they don't like with incredible ferocity. No sentence is too small or inconsequential not to argue about.[9]

Partly as a result, political campaigns have developed into hyper-condensed soap operas, as the media and rival political operations attempt to hype up each passing fluctuation in the polls, in order to create a sense of momentum (and in the journalists' case, to give themselves something new to write about). Consider the 2008 campaign in the US, in which Hillary Clinton went from presumptive favourite to hapless also-ran and back again, seemingly overnight – or, in 2012, the endless exhausting turmoil in the Republican field as candidates such as Michele Bachmann, Rick Perry or Herman Cain emerged to become front-runners before falling back just as quickly, or the oceans of media coverage given to Donald Trump in 2015 purely because he provided novelty, entertainment and outrage.

At this stage, wiser souls will point out that such froth is often far removed from the fundamentals. In the 2012 American elections, the press excitedly announced vast swings in the polls due to Barack Obama's lethargic performance in the first TV debate. Yet more level-headed pollsters such as Nate Silver – and the experts within the Obama campaign – could see that the fundamentals were unchanged: all that was happening was that wavering Republicans were firming up their allegiances.[10] And for all the excitement generated by the 'Milifandom' in the UK in 2015, the traditional fundamentals turned out to be far more important: voters thought David Cameron a more plausible prime minister than Ed Miliband, and the Tories better stewards of the economy than Labour.

The problem is that it takes a phenomenal amount of willpower, especially in the heat of a campaign, to ignore the moment-to-moment trends: the same Tories who spent the summer of 2015 explaining why their victory had always been inevitable spent the spring in a state of fretful, panicked uncertainty. The environment at campaign HQ is best summed up by Joe Trippi, who ran Howard Dean's grassroots-driven bid for the Democratic nomination in 2004. He described how, for those running the operation, 'the telephone is not a telephone. The telephone is a .357 magnum and every time you answer it and there's a reporter on the other end, what you're really doing is putting that gun to your head. Sure, most of the chambers are empty, but there's always at least one hollow-point in there, sometimes more.'[11]

It used to be that this mood would lift when the polls closed. But no more. The result has been an unfortunate dynamic in which politicians are not just hugely aware of the press, but hugely responsive to its demands. Barack Obama, for example, is probably the highest-profile advocate of the view that the acceleration of the news media is having a disconcerting – bordering on disastrous – effect on politics. He has made it clear that he believes, first, that the press are not doing their jobs properly, and second, that they are making it harder for him to do his, denouncing the media as fundamentally unserious and decrying its 'coarsening of our politics'.[12]

When, at a Q&A session, a young man asked him what had been most surprising about the actual work of being president, his mind leapt immediately to the dysfunction of the press: 'I've been surprised by how the news cycle here in Washington is focused on what happens this minute, as opposed to what needs to happen over the course of months, years. The 24-hour news cycle is just so lightning-fast and the attention span I think is so short that sometimes it's difficult to keep everybody focused on the long term.'[13]

Yet the paradox of Obama is that the same president who ordered his team to 'be about the long term' rather than focusing on coming out ahead that week could be on the phone to a senior staffer within 36 hours of his 2008 election victory complaining about a lack of momentum.[14] When things went well, Obama professed himself uninterested in the day-to-day flow of opinion polls. When they veered off the rails, he became as short-termist as anyone in his efforts to wrench them back on track.

The same pattern repeats itself across the world. A confidant of Gordon Brown's once told me that the only way to make him happy on a Saturday afternoon was to be the person who brought him the football scores – or the next day's headlines. His predecessor, Tony Blair, was so determined to pre-empt bad publicity that he and his press chief Alastair Campbell once phoned the Foreign Secretary, Robin Cook, while he was en route to the airport for his holidays. The *News of the World* had uncovered Cook's adultery;

Blair and Campbell insisted that the hapless minister must not only choose between his wife and his mistress, but do so before the first-edition deadline.[15]

Things have been made worse by the fact that stories, as we saw in the last chapter, can now come from anywhere and anyone. In October 2012 the UK chancellor, George Osborne, took a train back from his constituency. Although he had only a standard-class ticket, he decided to upgrade to first-class. All fairly straightforward – except that a journalist in a neighbouring carriage started gleefully tweeting about how the chancellor had tried to bag a first-class seat without a proper ticket, how his aide had refused an order that they move to sit with the plebs, and how the result had been a stand-up row with the ticket collector.

As journalist after journalist called the Treasury for more details, or just a comment, Twitter lit up with denunciations of the posho chancellor (estimated personal wealth: £4 million) who refused to slum it. To add to the confusion, a fake Twitter account appeared, purporting to be by one of Osborne's own staff. By the time the train reached Euston, the platform was surrounded (as the *Guardian* reported) by 'a feverish posse including Labour activists, the president and vice-president of the National Union of Students, and assorted press . . . officers from the Metropolitan Police's specialist response unit pored over train timetables to try to work out which service the chancellor was on, to make sure he was spirited away in safety'.[16]

Despite the use of Twitter itself by Osborne's staff to clamp down fast on the wilder rumours, 'the Twitter judgement had been made that this was a story', says one aide. 'It's this whole new world,' he adds, 'where you're constantly on camera, where people around you are on various combinations of social networks. You're basically never off camera or off microphone, because someone might be sitting in your carriage with 10,000 Twitter followers.'[17] On another occasion, for example, Team Osborne arrived back from the World Economic Summit in

Davos only to find that 'before we'd even got to the gate, someone on the flight had already called the *Daily Mail* complaining that the Chancellor was in business class'.

You could argue that, ultimately, such episodes fade from memory, but the need to be constantly wary keeps politicians and their staff in a state of permanent jitters. And it contributes hugely to what is the main problem with modern government: a lack of time, or willingness, to think things through.

GOVERNING IN A HURRY

'The pressures of the 24-hour media, to which I don't think any democratic government has adapted well, force one to be immediate and tactical,' Alastair Campbell has said. 'But government should always be strategic.'[18]

It is an observation echoed by his counterparts in the White House. 'Everything about the technology and the structure of the media today is designed to accelerate the flow of information,' Mike McCurry, Bill Clinton's press secretary, told Howard Rosenberg and Charles Feldman.[19] Marlin Fitzwater, who did the job for Ronald Reagan and George H. W. Bush, told the same authors that 'the world is operating at least three times faster than it did when I left office in 1992 . . . Everybody has to be quicker and better and smarter . . . but it often doesn't work that way because we don't have that quality of people'.[20]

Even as recently as the 1980s, the pressures were completely different – as was the way in which decisions were made. Stephen (now Lord) Sherbourne, once Margaret Thatcher's political secretary, points out that, unlike Campbell, her press guru Bernard Ingham viewed his job 'as defending the wicket rather than feeling the need to score any runs'.[21] Time could be taken to hone positions and think through policy. 'If there was a big speech on the Saturday, we'd work on it on Tuesday night, from say six till twelve, then the same on Thursday, then there'd be a line through the diary on the whole of Friday. It was a very, very prolonged suit-fitting – her way of getting away from the pressure of government to think things through. Nowadays, the idea of taking a whole day out would be crazy.'

The idea that the government is having to answer a few more phone calls and emails may not sound like much of a change. But in fact, it has had profound implications. The first is that it creates more government. 'Having a story a day isn't good enough now,' says one senior official in Downing Street. 'You have to have your morning story, then your afternoon story, then your evening story.'[22] No longer can governments have a 'health week' or 'welfare week', partly because attention spans are too short and partly because they simply can't produce enough new product to keep people interested. 'Media management becomes a lot more difficult, because you've got to try to feed this press machine,' says the official. 'The pressure is to generate new policy and new initiatives, which is just a really bad way to run a country. It creates bigger government.'

It also results in policies that are simultaneously half-baked and oversold. Just as the press has become increasingly sensationalist in its constant attempts to cut through to the audience, so has government. When making an announcement, admits the same No. 10 staffer, 'we need to make it as outlandish and outrageous as possible, make hyperbolic claims. There are so many press releases written by us, or by Labour, naming X as "the biggest reform in a generation". It's not at all, but we've got to give it this ridiculous spin to try to make sure it lands.'[23]

'We live in a world where consumers demand instant gratification,' says Mark Flanagan, who worked as a civil servant in Downing Street under the Blair, Brown and Coalition governments. 'If an issue arises in the media, the demand of politicians is "something must be done". This can lead to initiativitis, or government by sound bite, where politicians make instant promises on the hoof in a knee-jerk reaction. When that policy idea comes up against the inevitable practical problems of implementation, then the public get further disillusioned.'[24] Sir Humphrey Appleby in *Yes, Prime Minister* described the trap into which politicians fall: 'Something must be done. This is something. Therefore we must do it.'[25]

As a consequence, government too often moves from active to reactive – from thinking of new ideas to pandering to the latest popular trend. Public opinion, says Nadhim Zahawi, a Tory MP and founder of the polling firm YouGov, 'used to be like a fine Scotch whisky: sipped and savoured occasionally'.[26] Outside election season, Margaret Thatcher only received monthly updates on what voters thought, if that.[27] Now, governments swig from that bottle every day.

You could argue that this helps them respond instantly to voters' concerns. Yet all too often it leads to a focus on presentation over policy, and a willingness to back down in the face of noisy opposition (which itself is easier to put together in a more connected and less hierarchical age). As Zahawi says, 'polls can only tell you how you should communicate what you want to do. They can't tell you what you should do. Every policy creates a minority of losers, yet it's always the losers who are best organised and most vocal, particularly in an online arena.'

That outcry can also, thanks to the great acceleration, arise from the most unexpected places: witness the fact that a board for discussing Japanese pop culture, 4chan, gave birth to Anonymous,[28] or that a movement to boycott American beef was given much of its impetus on the fan sites of South Korean pop groups.[29]

This atmosphere also leads to a greater temptation for politicians to be more partisan, so as to whip up online support (or counter their enemies' hostility). As Obama's communications director, Dan Pfeiffer, ruefully admitted in a memo to the president on the failures of his first term: 'We made a strategic choice that we were gonna portray you, because you are, as reasonable. But in today's media environment, there is no caucus for reasonableness, with the possible exception of David Brooks.'[30]

Perhaps the chief impact of a faster news agenda is that it makes it harder and harder for leaders to pursue their own agenda, rather than being at the mercy of events. Peter Hennessy, one of Britain's leading constitutional historians, has described the prime minister's job in terms of the waves and particles of quantum physics: 'The waves are the wider sweeps of events, technological changes, shifts in the global political economy. The particles are the problems and pin-pricks, many unexpected, several of short duration, but great absorbers of mental and physical energy while they last, that so often take up so much of a prime minister's day.'[31] Barack Obama recognised something similar before taking office, acknowledging that 80 per cent of his job would be taken up by reacting to events, leaving only 20 per cent of his time for pushing forward his agenda.[32] The *Washington Post* columnist Dana Millbank complained that 'our national dialogue has become a series of one-act plays: each runs for a week or two, the critics volunteer their reviews of the president's performance, and then it closes just as quickly'.[33]

The question – perhaps the key question – is whether this is merely a drag on good government, or a fatal obstacle. Some are convinced of the latter. Hodding Carter, a veteran of his namesake's administration, told Rosenberg and Feldman that 'the present media rules of the game destroy the ability of government to move at all rationally'.[34] I was once briefed by an aide to the French government, who had developed a 14-stage chart, full of looping lines and arrows, that purported to show how a story developed on Twitter, and where and when to inject a ministerial comment. As an attempt to impose order on chaos, it was hard to beat.

This leads to an awful paradox: that a world which speed has rendered more fragile, more prone to sudden disruptive changes both positive and negative, is also one in which government is less and less able to cope with crises. In Tony Blair's memoirs he claims that the accelerated agenda means government today needs to make decisions at the speed of light, compared with the speed of sound in days gone by.[35] Yet in the same book, there is a frequent sense of quite how paralysing it can be to be in the heart of the storm. During the foot-and-mouth crisis in farming (though he could also be writing about the petrol strike that briefly paralysed the country), Blair was most struck by 'the utter incapacity of the normal system to deal with abnormal challenges . . . when I look back and reread the papers, reminding myself of the sheer horror, depth and scale of the crisis, it is a total miracle we came through it'.[36]

'The modern media environment doesn't allow for considered decision-making in the way it should,' says one Downing Street insider.

> Let's say something breaks at midday – the press want a decision: 'Are you going to sack the guy or not?' If you haven't made a decision, it's 'PM dithers over giving minister the sack'. A lot of our day is spent managing these stories, and deciding semi-instantly on the fate of a policy or someone's career, which I think makes for quite corrosive government. And when there's a crisis, it takes over the place. There's no time to think any more. There's no time to make a properly considered decision.[37]

As long ago as the Crimean War, generals were complaining about being micro-managed by distant politicians, thanks to a newfangled invention called the telegraph.[38] In the negotiations over the Versailles treaty, leaks of the progress being made appeared in the papers, which prompted statements from governments, which toughened the stances of their negotiators.[39] Yet in today's turbocharged world, it sometimes feels as though foreign policy prescriptions are being delivered by the talking heads of cable news not in accordance with what would be wisest, but what would drive the story forward.

In a profile of Barack Obama for *Vanity Fair*, Michael Lewis described how Republicans such as Newt Gingrich moved effortlessly from calling for intervention in Libya to castigating the president when he finally acted. 'The tone of the news coverage shifted dramatically . . . one day it was "Why aren't you doing anything?" The next it was "What have you gotten us into?"'[40]

George Kennan, the architect of containment theory during the Cold War, observed of US intervention in Somalia (in the form of the events that inspired the film *Black Hawk Down*): 'If American policy . . . is to be controlled by popular emotional impulses, and particularly ones invoked by the commercial television industry, then there is no place . . . for what have traditionally been regarded as the responsible deliberative organs of our government.'[41] Henry Kissinger, another foreign policy veteran, has voiced similar concerns that a 'mad consensus' will drive the world, with leaders unwilling to stand up to popular feeling.[42] The *New York Times* columnist Thomas Friedman describes this as a shift from representative democracy to something closer to *American Idol*, with policy being dictated by instant plebiscites of an emotionally volatile population.[43]

How bad could the problem get? In a rather alarming thought experiment for their book *No Time to Think*, Rosenberg and Feldman asked Ted Sorensen, speechwriter for President Kennedy, to imagine the Cuban Missile Crisis taking place in the 21st century. Sorensen considered that:

> In all likelihood, today's media pressure would have made it impossible for the Kennedy team to keep confidential for one week the fact that we knew Soviet missiles were in Cuba. In all likelihood that would have meant public panic, and congressional pressure would have required the president and ExCOMM to decide on its response a week earlier. In all likelihood, that would have meant our selecting everyone's initial first choice, an air strike against the missiles and related targets, which, in all likelihood, would have required,

> according to the Pentagon, a follow-up invasion and occupation of Cuba. And in all likelihood – in as much as we discovered that Soviet troops in Cuba were equipped with both tactical nuclear missiles and the authority to use them against any USA attack – the result would have been a nuclear war and the destruction of the world.[44]

The potential destruction of humanity is a heavy burden to place on the shoulders of an accelerated news agenda. But there is no doubt that the feedback loop between press and politicians (and, indeed, public and politicians) has vastly accelerated, with a concomitant weakening of politicians' ability to act as deliberative representatives – to apply the filter of their own wisdom and experience to the public's instant demands.

This is not only true of democracies. One of the main problems in resolving the tensions that frequently flare up in the Pacific is the 'clickbait nationalism' fostered by the Chinese government, in which jingoistic bloggers whip the population into such a patriotic frenzy (especially if Japan is on the other side of the argument) that it becomes impossible for Beijing to back down.[45]

Fortunately, there are some consolations. First, when it comes to the crunch, politicians are still capable of acting decisively to stave off disaster. Think of the way catastrophe was eventually averted during the financial crisis, or the fact that, for all the pressures of office, President Obama was still able to take several months to determine whether to reinforce American troops in Afghanistan in his first term (and later to plot the raid that killed Osama bin Laden while simultaneously honing his gags for the White House Correspondents' Dinner). Those at the highest level of the British government point out that, despite all the media pressure, they have still been able to accomplish big, difficult, long-term things – reforming the state pension system, overhauling welfare and education – whose intended benefits could take years, if not decades, to be felt.

And there are also positive consequences of acceleration: for example, that errors are uncovered more quickly. It is not just that there is always someone there with a video camera if your candidate uses a racial slur, or a record of everything embarrassing and contradictory that you have said. It is that every statement a politician makes is subject to instant fact-checking.

In Britain, this transformation can be seen most clearly during the annual Budget. This is at the heart of the political calendar, the moment at which the chancellor charts the course of the economy for the year to come. So important is its news management that under Gordon Brown the Treasury would actually write the alerts being sent out by Reuters. 'We'd have this ultra-embargoed conversation,' says Damian McBride, then head of communications at the Treasury, 'where we'd say "When he gets to this line in the speech, you are allowed to snap this", which would say "GDP forecast for the next five years". We had incredibly strict rules, but it saved them ten seconds writing it out and made sure they didn't get something wrong.'[46]

For Reuters, that ten seconds was crucial in stealing a march on their rivals – speed in action. But the process as a whole still retained something of its old stateliness. During the parliamentary debate, Brown's facts usually went largely unchallenged by the Opposition; journalists would instead gather at the Institute for Fiscal Studies the next morning to hear its experts pick apart the detail.

Yet with the fact-checking process now crowd-sourced via Twitter, any errors will emerge more quickly than ever. George Osborne's 2012 Budget statement was certainly given less than the proper amount of attention during its preparation, but it was still startling that the 'omnishambles Budget', as it became known, crumbled so fast.

'Less than an hour after George had sat down, I was in the middle of briefing the lobby, and people were already looking at their phones, and "granny tax" [one of the least popular measures] was starting to trend on Twitter, and I could feel we were losing control of the story,' confesses one adviser.[47] It is not just errors – the decentralised media environment that enables stories to emerge from anywhere also means that established narratives, facts and policies can be challenged more rapidly and successfully.

MEN OF DESTINY

So far, we have focused on what the acceleration of the news agenda has done to politics, chiefly by reshaping it around the mayfly enthusiasms of voters and journalists. But there is another aspect to its effects – namely, what it has done to politicians. In an instant age, what qualities does it take to get to the top of the political tree, and then to do a decent job when you get there?

The first point is perhaps the simplest: the pressures described above have made the job of a leader harder on a basic physiological level. We saw earlier how the information age can contribute to stress, lack of sleep and disrupted attention spans – and the effect in an already stressful profession such as politics has been even more intense.

Above all, therefore, a modern leader needs to have the right temperament to cope with such pressures. Indeed, as the demands of the news agenda mount, you can make a good argument that this has become the single greatest determinant of their ultimate success. 'The best leaders now,' says one senior Tory, 'are the ones who can make a quick decision and say: "Right, this is what we're doing, I'm going to stick to it." That's why Gordon Brown was a disaster – because he could never take a decision.'[48]

According to no less an authority than Tony Blair, time management is among a leader's pre-eminent tasks. Putting a meeting in his prime ministerial diary that he did not genuinely need, even when he had warmly assured that individual that it could not happen soon enough, was referred to jokingly as an 'S.O.' – 'sackable offence'.[49]

Brown, whatever his flaws, was a politician of immense experience, with an able team around him. And he was more aware than anyone of the turbulent nature of modern politics. The problem was that he had a personality uniquely unsuited to Downing Street.

Damian McBride, a former Brown aide who is largely sympathetic to his old boss, admits that 'Gordon had a terrible time-management problem'.[50] As chancellor, he had largely controlled his own diary, but as prime minister there were endless duties, such as meetings with ambassadors, that needed to be squeezed in. 'He found it really, really difficult,' says McBride. 'He'd then come tearing around in the free time between one meeting and another saying: "What's going on? What's going on?" And whatever was on the TV at the time, you'd have to explain to him why that was the big story on the telly, what we were doing about it, what our line was, whether he needed to do anything about it.'

Things were made worse by Brown's redesign of No. 10, which saw him and his team govern from a vast open-plan 'war room', intended to give them constant command of events. 'On one big wall was Sky News, and on the opposite wall was BBC News 24,' recalls Lord Adonis, the then transport secretary. 'The effect was constant interaction with the world outside – especially since the cameras would be literally outside No. 10. There was a constant sense of the need for media management.'[51]

What developed was an addiction to action for action's sake. To give the impression of decisive leadership, Brown took to convening Cobra – the government's emergency committee – for even the most minor emergencies. Alistair Darling, his chancellor, writes in his memoirs of the 'hopeless' management of Brown's time leading to 'a permanent air of chaos and crisis' in Downing Street, with endless meetings but no decisions made.[52] In the wake of the financial crisis, Brown 'wanted to make fresh announcements seemingly on a weekly basis . . . I remember one occasion when, as we were about to do a series of media interviews, [fellow minister] Yvette

Cooper and I found that we could not remember all of the things the government had announced.'[53] 'The foundations were so weak under Gordon that the crises threatened to finish you off,' says Mark Flanagan, another member of that Downing Street team.[54]

So a good leader needs an ability to make snap decisions, to manage their diary effectively, and to create a distance from the instant judgements of the media. But they also need something rather more basic: a rigorous exercise regime.

It was startling to witness first-hand the extent to which, over the first two years of the Coalition being formed in Britain, those in power piled on the pounds. The demands of the job were such that late nights and fast food had replaced healthy eating and work–life balance. In the US, Dan Pfeiffer, a senior Obama aide, became so stressed that he suffered a series of mini-strokes.[55] This, rather than a public preference for youthfulness, may be one reason why leading politicians are getting younger and younger: because by the time they reach maturity, they're exhausted.

'I think the accelerated agenda hollows out leaders faster,' says John McTernan, a former Blair aide. 'If you look at Thatcher, we'll never have a prime minister who does the job for as long.'[56] Lord Adonis agrees that 'the pace of politics is so more much more intense than it was. You need to be very physically fit in a way that you didn't need to be. And people do seem to burn out sooner than they did in a previous generation – we definitely don't have large numbers staying on into their sixties, and carrying on doing politics.'[57] Not least because any signs of ill health will be picked over by the media. As Lord Adonis points out, it is simply inconceivable that a prime minister today could suffer a stroke, as Winston Churchill did in June 1953, and have the whole thing hushed up.

> He did go through a period of not attending the House of Commons, but they had a very long summer recess and the only thing he had to do was to appear at Conservative Party conference. The

decision on whether or not he was going to remain prime minister was dictated by his ability to deliver that speech. And he was just up to doing it, they just managed to get him fit enough. There's no way you can imagine that now. Can you imagine David Cameron not being in the media for a week, let alone a month, let alone four months?

JUDGING BY APPEARANCES

One of the main problems in politics, therefore, is that the kind of leaders an accelerated environment demands – calm, level-headed, able to cope with pressure, manage large teams and deliver on detailed policy promises – are not necessarily those which the media environment favours.

Again, this is not necessarily a new problem: the public have always been suckers for pretty faces. What Malcolm Gladwell refers to as the 'blink test' – the instant snap judgement of the voters – is a powerful tool, but often a wayward one.[58] He cites the example of Warren G. Harding, a disastrous and dim-witted president elected largely because he was 'irresistibly distinguished-looking'.[59] But there are plenty of other examples. In an experiment by the Princeton psychologists Alexander Todorov and Charles Ballew, members of the public were shown images of competing candidates for just a quarter of a second, and asked to guess which one looked more competent. The chosen candidates turned out to have won 70 per cent of real-life races.[60] Likewise, a study of the 1972 federal elections in Canada found that the more handsome candidates received two and a half times the number of votes.[61]

An impressive appearance is not, of course, the be-all and end-all: if it were, Mitt Romney would have won the Republican nomination in 2008, rather than John McCain. But we still place an inordinate amount of weight on how convincing candidates look on TV. Yet as Rosenberg and Feldman argue, the qualities that win the voters' approval on television may not necessarily be those that make a good president or prime minister. Political debates, for example, 'celebrate hair-trigger ad libs and quick, agile wit . . . along with flashy TV skills, they reward that familiar bugaboo, speed . . . the candidate who weighs an answer thoughtfully before responding – a trait we should value in a president – most often appears indecisive, flustered, out of touch, a musty old soul worthier of assisted living than the Oval Office.'[62]

In Britain, the 2010 elections were the first to feature televised debates between the candidates – and the reaction appeared to bear this out. Nick Clegg, leader of the Liberal Democrats, went from nobody to national pin-up not because of the strength of his policies, but because of his charming habit of looking straight into the camera and addressing questioners by name. After participating in a similar debate between the parties' financial spokesmen, Labour's Alistair Darling said he 'was struck by how much people commented on the style and demeanour of the candidates. When I would gently ask what they thought of what we had said, very little had got through.'[63]

Yet it's not just debates any more. One of the most interesting political developments recently has been the sudden ubiquity of the 'selfie', a transition that one of my former colleagues compared to the arrival of the talkies in the cinema. All of a sudden, politicians have had to master a completely new skill, namely the ability to look relaxed and friendly while hugging a voter and staring into their mobile phone. Appear too awkward, and the result will whip round social media at light speed.

Some politicians have actually turned selfies into a cornerstone of their campaign strategy. John Key, prime minister of New Zealand, took more than 20,000 on the campaign trail in 2014. His team claimed that each would be shared more than 100 times – enough, in theory, to reach half of the electorate, achieving maximum exposure without needing to answer tricky policy questions or do anything other than look vaguely genial.[64]

All of this – catering to the instant, superficial side of our accelerated culture – is a recipe for bland, plausible candidates. It is also a recipe for public boredom and alienation. When working as a Republican strategist, Roger Ailes, who later founded Fox News, pointed out: 'If you have two guys on a stage and one guy says "I have a solution to the Middle East problem", and the other guy falls into the orchestra pit, who do you think is going to be on the evening news?'[65] The result is a style of political campaigning in which it is far, far more important to avoid gaffes than to say anything interesting – hence the fact that politicians only interact with voters on the campaign trail under the most carefully controlled circumstances.

You might think that all this superficiality would attract voters' scorn rather than support – and you'd be right. Professor Steve Barnett of the University of Westminster argues that the hierarchy of social deference has flipped over: instead of trusting leaders, elders and experts over celebrities, friends and random acquaintances, we now view them with increasing contempt.[66] One of the more interesting developments in recent years, indeed, is the increasing profile and popularity of deliberately unpolished politicians – Nigel Farage, Donald Trump, Beppe Grillo, Jeremy Corbyn – who set themselves against the established elite, who seem to speak as tribunes for the everyday man.

Of course, such populist positioning is nothing new: it was invented by the ancient Greeks along with democracy. But it has been given new life by the ease with which new movements can froth up online, by our impatience with governments that continuously fail to deliver with the speed and efficiency we expect – and by the fact that the great acceleration has polarised the economy between haves and have-nots, with entire communities or professions left behind by the remorseless competition that is a feature of the modern economy. The Corbyn ascendancy within the Labour Party, while enabled by the peculiar features of the party's system of leadership elections, was fundamentally driven by a revolt on the left among those who felt that the political system, and the solutions it has championed for decades, simply isn't delivering.

RIDING THE TIGER

Our leaders are not, whatever we may think, stupid people. They are as aware of this process as anyone. Indeed, one of the most interesting, and under-appreciated, aspects of modern politics is the extent to which politicians are deliberately shaping their media strategy around the demands of an accelerated age and the prejudices of an ever more fickle public.

One thing that is widely accepted, for example, is that the public are growing tired of politicians more quickly. Gordon Brown had a theory, recalls Damian McBride, that:

> you've got seven years. That's about how long you can be exposed to the public before anyone who's ever going to warm to you will have done so already, and lots of those that have will be starting to go off you, and those that never liked you are now outright hostile because they're thinking: 'When will you ever get off my telly?' Gordon was very careful to ration his public appearances because of that while he was chancellor. He wanted to be on the news, but he wasn't chasing it every single day.[67]

John McTernan, a former adviser to Julia Gillard of Australia as well as to Tony Blair, agrees that we use up and spit out our leaders much more quickly: 'We've got a throwaway culture, not just in goods but in ideas and ideals, and politicians.'[68]

The problem is that those same politicians are in constant demand, to comment on and push forward the stories of the day.

'The media decides that *this* is the story, *this* is occupying everyone's attention, and people start saying, "We haven't heard anything from Downing Street. . .",' says McBride. 'They need to update that story in real time and you are part of that story and how it unfolds.'

Let's not blame journalists entirely, however: politicians are invariably suckers for publicity – and public approval. The knee-jerk option is therefore for them to insert themselves into the news agenda at any opportunity. Cameron pronounces on racism in football.[69] Blair orders his home secretary to investigate the case of Deirdre Rachid, an unjustly imprisoned character on *Coronation Street*.[70]

This strategy, however, is not as ridiculous as it may seem. Barack Obama, for example, sits at the head of a monumentally effective, monumentally ruthless communications machine, one which pumps out, on Twitter, Facebook and every other channel it can access, its own product: Brand Obama. In his biography of Obama, *The Promise*, Jonathan Alter explains the strategy:

> The most common media criticism of Obama [in his first year in the White House] . . . was that he was overexposed. He granted scores of interviews, issued op-ed pieces, taped weekly radio addresses and webcasts, popped into the White House briefing room for daytime press conferences, appeared on late-night TV, and held a televised town hall meeting outside of Washington on average every couple of weeks. 'You're the president, not a rerun of *Law and Order*,' Bill Maher joked.[71]

But, says Alter, 'Most of the time Obama was overexposed only to those who regularly watched Fox News, CNN, MSNBC, PBS, Comedy Central or CNBC or listened to NPR. That sounded like a big audience, but in total it amounted to only around 10 per cent of the electorate who voted in general elections . . . the other 110 million voters who were largely uninterested in politics were not being force-fed very often.' He adds: 'Even when his job approval ratings drooped, his personal ratings stayed healthy . . . the Obama White House was the biggest, splashiest reality show ever.'[72]

David Cameron takes much the same tack. As his party's best spokesman and salesman, he constantly feels the urge to inject himself into the conversation. Perhaps that shortens his shelf life, but it also helps his team to fill the news void and communicate with different audiences, instead of coming up with endless costly and counterproductive policy gimmicks.

Another variant on this approach, adopted by Obama and Nicolas Sarkozy, is to try to overwhelm the media, and your opponents, with your speed. The idea is to push the government machine into overdrive in the hope that your goals will be reached before the engine stalls. Washington in the early days of Obama, says Alter, 'was a blur of activity . . . no one could possibly keep up with the Niagara of news coming out of the administration.'[73] The problem is that such an approach also breeds errors: getting something – anything – done takes precedence over making sure it is the right thing.

There is, however, a possible alternative, which no mainstream politician has been brave enough to try. While there are many problems with online discourse – particularly its tendency to engage in witch hunts and public shaming – it also tends to privilege more positive qualities: honesty, openness, wit, humour, a genuine human voice. These are also what we claim to be crying out for in our increasingly stage-managed politicians. So what if more politicians surfed the online wave, rather than letting it wash over them?

'In response to the transparent era of Twitter, Instagram and YouTube, campaigns can either run and hide, or they can dive in headfirst,' argued CNN's Peter Hamby in his autopsy of the 2012 presidential election campaign. 'Trying to navigate a middle ground is a path fraught with danger.'[74] As Chuck Todd, a senior US news journalist, told Hamby: 'Some candidate [in 2016] is going to say:

"I'm going to make this my advantage. I'm going to take the fact that the news cycle is 24 one-hour news cycles. So why not be totally unfiltered . . . and just say everything is on the record, everything is open-sourced. The first candidate that cracks that code and does it will get rewarded.'[75] The attraction of candidates such as Jeremy Corbyn – or even Donald Trump – is precisely this: that they appear to be their own men, to speak their minds and offer solutions uncontaminated by the shabby compromises of Westminster or Washington.

How long will it be before, to prove their bona fides, a politician genuinely 'goes clear'? In Dave Eggers's techno-parable *The Circle*, politicians end up wearing cameras that broadcast their every conversation, 24/7, in order to demonstrate their integrity.[76] It may sound ludicrous now, but as cynicism mounts about politics and politicians, who's to say it won't catch on?

GETTING BETTER GOVERNMENT

The situation can be summed up as follows. In a flashy, charisma-driven age, we vote for politicians who promise freshness and novelty, who run against business as usual and the shabby compromises of power. The problem is that once these politicians get into power, they prove unable to meet those promises – indeed, they degenerate rapidly into greying, paunchy pastiches of their younger, more promising selves. Our disillusionment is all the more severe for having had our hopes raised in the first place – meaning that we will be less inclined to trust any politicians at all.

And, as ever, this process is heightened by the other effects of the great acceleration. As mentioned above, the growth of the anti-politics mood – and of movements such as UKIP in Britain, or the Front National in France – can be pretty much directly interpreted as a reaction to an accelerated, globalised world, in which the divisions between haves and have-nots grow ever sharper. As society becomes more unequal, it is an iron law that it becomes more politically polarised, which makes consensus harder to achieve.[77] Also, in an individualistic age, we are ever less willing to subordinate our identities to those of the traditional political parties – which helps explain why in 2012, only four of the 34 members of the OECD had governments with an absolute majority.[78]

Matthew Taylor, a former senior adviser in the Blair government, is now the chief executive of the Royal Society of Arts, and one of Britain's foremost students of political disillusionment. He believes that the work of the sociologist Avner Offer holds a valuable clue to what's going on.

Offer argued that recent Western history consists of a more or less successful effort to rid ourselves of what he calls 'commitment devices' – social structures such as family, friendships or religious belief that prevent us from indulging our most immediate desires. 'In the Sixties and Seventies,' says Taylor, 'as we became very affluent, we decided that we didn't need these things any more – we didn't need marriage, we didn't need trade unions, and why should we have any limitations on borrowing? So we basically dismantled all these things, on the assumption that in an age of affluence, we can do whatever the hell we want. [But] once we start to dismantle these devices, we start to make disastrous decisions.'[79]

The effect of the great acceleration, argues Taylor, has been to give these centrifugal forces a huge push.

> Technology and pace are things which undermine hierarchy, which undermine the kind of routine relationships that underpin solidarity, most fundamentally through mobility. Pace suits individuals, individualism in its most rampant form is 'I want what I want and I want it now'. But it further undermines hierarchies, because hierarchies move slowly, because of the amount of time it takes information to move from the top to the bottom.

There are few things more obviously hierarchical, of course, than government. So, says Taylor, 'because it is unable to respond to the pace of life, what you get is a longing for alternative forms of authority – for populist, charismatic forms of leadership, for example.'

There is a fundamental problem here, in other words, that goes beyond politics and to the heart of what government is. In an accelerated age, the state is still set up to be slow – partly to foster time for deliberation, but mostly because of the dead hand of bureaucracy, inertia and, frequently, incompetence.

In his memoirs, Tony Blair complained of having to pretend to be normal on the campaign trail: going to the shops, buying ice cream with Gordon Brown in order to persuade voters that they did not actually loathe each other. 'Before the rounds of interviews anywhere near election time,' he moaned, 'I would have to go through a list of the price of everyday things like a pint of milk, a pound of butter, a shoulder of lamb. Bread used to produce lengthy debate about which type of loaf, white or brown, nothing too wholemeal, nothing too unhealthy.'[80]

Yet what Blair failed to realise was that, even with such periodic reality checks, he was living in a bubble. The *Wired* journalist Ben Hammersley points out:

> With good luck and a following wind, you might expect to be in Downing Street for eight solid years. But the technology policies you put forward in your first year in office will be based around technology that will be laughably obsolete by the time you leave. The mobile phone on which you took congratulatory calls during election night will have been replaced by something 16 times as powerful.[81]

Blair only sent his first text message after leaving office in 2007; Alastair Campbell (now a prolific blogger and tweeter) never typed an email in Downing Street.[82] And when the Tories arrived in No.

10 in 2010, they found that the 'autocomplete' function on email addresses had actually been turned off. When one frustrated staffer asked why, he was told – perhaps mischievously – that Gordon Brown had been firing off an angry, expletive-fuelled email to Wendy Alexander, his election co-ordinator in Scotland – and had sent it to Rupert Murdoch's then wife, Wendi Deng, instead.[83]

A society based on instant gratification demands the same speed and responsiveness from government as it does from any other service. But not only are our political systems under more pressure than ever before, but government's relative imperviousness to the forces of technological change means that a fault line opens up between what we expect and what it can deliver. Thinkers such as Al Gore or Lawrence Lessig, the digital rights activist, are vociferous in their view that government is simply not equipped to tackle the great challenges of our age.[84]

For one thing, the great acceleration itself, by making it easier for people to organise, has made it ever easier to throw grit in the wheels of government. In the words of David Frum, a columnist and former White House speechwriter:

> There's a lot of technological triumphalism about how [the internet] can be used to improve democracy. But in the end what seems to have happened is that it's empowered angry and highly motivated minorities, and empowered them to slow down the system. Getting things done seems to go slower and slower every decade. How long does it take to build a highway? How long does it take to build a bridge? How long does it take to get a presidential nominee through the Senate? Those things are all slowing down.[85]

Some people have argued that the solution to this is more transparency – perhaps along the lines of the Eggers model above. Frum is sceptical: 'We keep thinking that the way to inspire more confidence in government is more transparency. But the process of government is often not very inspiring – and now we all have glass windows around the sausage factory. So people have lost confidence in government.' Archon Fung, of the Kennedy School of Government, makes a similar case that while transparency lets you catch mistakes, it also breeds cynicism about the process. It's 'like creating a big Amazon rating system for government that only allows one- or two-star ratings'.[86]

A related problem is that, now that everything is public, and can be responded to instantly, government becomes fatally terrified of getting things wrong, and of the criticism that will follow. One of the reasons why no one has been prosecuted for causing the financial crisis is, as the US journalist Matt Taibbi has shown, a deliberate decision by the Attorney General's office under Obama not to prosecute any cases that it might lose, for fear of the negative publicity: so the banks paid with fines and apologies rather than jail time.[87]

In such an environment, politicians also become reluctant to take long-term risks if it may result in short-term punishment. Actually, they needn't be: Elke Weber at Columbia University has found that within six or nine months of unpopular measures like smoking bans or carbon taxes being introduced, voters had accustomed themselves to the status quo and entirely forgotten their objections.[88] But it's hard to tell that to a politician staring at flatlining polls and angry protests. The result is what the political theorist Francis Fukuyama refers to as a 'vetocracy', in which blocking things is far, far easier than doing them.[89]

There are also the effects of the permanent campaign to contend with. Frum argues that, with the need to fund and run a permanent re-election operation, 'members of Congress have to spend an enormous portion of their time, a quarter, maybe more, raising money. And then they're all now expected to go home at weekends, thanks to improved travel technology, where they do more fundraising, and more meetings with activists, and yet more fundraising.'[90]

Al Gore refers to this as 'quarterly democracy' – a spin on 'quarterly capitalism' which sees meeting fundraising targets every four months become the be-all and end-all of politics.[91] Eighty per cent of this money, he notes, goes on television adverts – presenting a huge barrier to entry for newcomers, and ensuring an 'obscene dominance of decision-making' by wealthy contributors. 'Not since the 1890s,' he claims, 'has US government decision-making been as feeble, dysfunctional and servile to corporate interests as it is now.'[92]

As a result of all this, there is an increasing sense that government as a whole is no longer fit for purpose. In Silicon Valley, in particular, there is widespread concern that its ossified structures are becoming a crippling drag on growth and innovation. George Packer, writing in the *New Yorker*, cites the case of a Silicon Valley employee who refused to take time out to listen to Barack Obama speak in his office: 'I'm making more of a difference than anybody in government could possibly make,' he insisted.[93]

Generally, Packer maintains that the technology industry sees government as 'slow, staffed by mediocrities [and] ridden with obsolete rules and inefficiencies'.[94] For example, the patent system, which in theory is meant to foster innovation, is so clunky, and so open to abuse, that one lawyer and his five-year-old son were able to patent a 'method of swinging on a swing', 'in which a user positioned on a standard swing suspended by two chains from a horizontal tree branch induces side-to-side motion by pulling alternately on one chain and then the other'.[95]

In its own survey of innovation, the *Economist* concluded that there is actually as much of it around as ever – it is just that our governing institutions are too rigid to absorb it.[96] Indeed, Eric Schmidt, the executive chairman of Google, and his colleague Jared Cohen lament in *The New Digital Age* that 'the innovative edge that is the hallmark of the American technology sector is largely walled off from the country's military by an anarchic and byzantine acquisitions system . . . Big defence projects languish in the prototype stage, over budget and behind schedule, while today's commercial technologies and products are conceived of, built and brought to market in volume in record time.'[97] They cite the example of helicopter pilots in Afghanistan strapping GPS-enabled iPads to their knees, using the mapping systems to circumvent the cumbersome in-built systems.[98]

'There's a balance between innovation, entrepreneurship and regulation,' Schmidt told me when I interviewed him about the book. 'You can imagine these systems becoming so over-regulated that it's not possible to do innovation. With Google Glass, people are already calling for its regulation and they haven't even used them.'[99] Conflicts between software companies keen to disrupt society and governments keen to protect the status quo are becoming increasingly common: the UK has taken on the digital giants over terrorism and privacy;[100] the EU is mounting a sustained attack on Google's monopoly in search;[101] and New York recently went to war with Uber.[102] (In a possible taste of things to come, it lost spectacularly.)

REINVENTING GOVERNMENT

So how can government deal with all these pressures – and the opportunities and dangers of disruption? One conventional prescription to resolve these problems is for politicians to look to the long term, to beef up their 'horizon-scanning' function. As the former Blair aide Geoff Mulgan has said, many governments 'live with their eyes on the rear-view mirror, refighting ancient battles and reigniting ancient enmities'.[103]

How much better, instead, to develop a culture more akin to that of the Iroquois, whose Great Law, as cited by Philip Zimbardo, stated: 'In every deliberation, we must consider the impact on the seventh generation . . . even if it requires having skin as thick as the bark of a pine.'[104]

But it is all very well to talk about making decisions for the long term. The trouble is that when the world is moving so fast, even the best planners can be blindsided. In its *World Economic Outlook* for 2007, the IMF cheerfully proclaimed that 'overall risks . . . seem less threatening than six months ago'.[105] We know what happened next. Or take the HS2 high-speed rail line in Britain: it has taken decades to plan, and will take decades more to build, by which time regular high-speed rail may be laughably outdated in the face of maglev trains, mega-fast 'hyperloop' bullet trains or simple old-fashioned ultra-fast broadband.

Lord Browne of Madingley, the former head of BP, now serves as the government's lead non-executive director, charged with bringing business savvy to the affairs of government. His mantra, he says, is not to look more than a decade ahead, but instead to focus on a few immediate priorities and devote all your resources and attention to driving them through.[106] Others suggest making smaller, more flexible bets that can be adjusted in the light of changing events.

When it comes to technology in particular, there is a host of ways in which the disruptive, fast-moving, anti-hierarchical spirit of the great acceleration could be employed to make government – and politics – work better and faster. For example, one Downing Street official points out that changes in technology 'give politicians an opportunity to engage with people in a way they never would have five or ten or fifteen years ago': they are no longer reliant on the media to make their case.[107]

And while the internet has enabled candidates to bypass the media (most effectively with Barack Obama's incredibly disciplined get-out-the-vote effort in 2012), it has also enabled them to draw their supporters into the process. Joe Trippi says of his work on Howard Dean's campaign in 2004: 'It became obvious pretty quickly that a couple of dozen sleep-deprived political junkies in our corner offices . . . couldn't possibly match the brainpower and resourcefulness of six hundred thousand Americans.'[108]

On a more provincial level, British MPs have used the web to compile email lists of a significant fraction of their constituents, enabling them to respond more immediately and directly to their concerns. Sometimes, technology has even taken politicians out of the process altogether: UKIP's first MP, Douglas Carswell, notes how appeals from parents with disabled children, previously a huge part of his workload, dried up as they began to organise online, providing advice to each other on how to ensure that the system gave their children what they needed.[109]

Government will, by its nature, always move more slowly than the rest of society. First, there's the inertia of any bureaucracy to contend with: for example, if terrorists made sure to switch to the very latest form of communication every couple of weeks or so, it would be vastly harder to track them – because by the time the requests and requisitions had worked their way up and down the intelligence hierarchy, the bad guys would already have moved on.

There's also the more basic fact that it's impossible to be jazzily disruptive when you're also accountable to the voters, and to innumerable regulations, directives and treaties – after all, 'move fast and break things' doesn't really work when the thing you're breaking could be people's tax records or welfare payments. One friend was horrified to see the head of the civil service using his Twitter account (itself a sign of the times) to promote a competition offering an innovative civil servant a chance to study at Singularity University, the centre set up by Ray Kurzweil to spread the doctrine of exponential acceleration.[110] What its sponsors saw as getting with the spirit of the age, he saw as suborning government to the ideologies of Silicon Valley.

Yet whatever your views on the ideal role of government, it's hard to argue that the current system is really the best way to supervise the spending of between a quarter and half of our gross national product. Douglas Carswell quotes the economist Kenneth

Boulding, who wrote that 'the larger and more authoritarian the organisation, the better the chance that its top decision-makers will be operating in purely imaginary worlds'.[111] That is not a bad way of describing modern government.

Tony Blair, after more than a decade in No. 10, concluded that 'the old infrastructure of policy papers submitted by civil servants to Cabinet, who then debate and decide with the prime minister as benevolent chairman, is not suitable in responding to the demands of a fast-changing world or an even faster-changing political landscape'.[112] A recent report by the think-tank Reform highlighted the 'red box' system as a particular problem in Britain. Under this centuries-old practice, ministers are handed dozens of files by civil servants and asked to adjudicate on them. As one of the report's anonymous sources said, this is equivalent to a system in which 'the CEO will literally make every decision in the entire company, and what we'll do is give him quite long pieces of paper to take home, before he goes to bed, to take with him, to have a glass of wine, scribble on and bring back in'.[113]

The problem is even worse when it comes to relations between different departments, which communicate formally via letters between ministers – roughly 15 per person per day, by one estimate. The crowning absurdity came in March 2013 when Michael Fallon, a business minister, was given added responsibility for energy, which was housed in another department. On his very first day, 'Business Fallon' had to sign a pre-written letter to 'Energy Fallon', asking for a change in policy – only for 'Energy Fallon' to politely write back telling himself to get lost.[114]

In a strange way, however, it is the very fact that government's problems are so significant that makes the potential benefits so vast. Clayton Christensen and others have written powerfully about the opportunities in teaching and healthcare that could come from embracing innovation and technology.[115] But there is much more low-hanging fruit out there.

For example, when the Conservatives took office as part of the Coalition, they were determined to make major changes to one area in particular: digital government. Back in 2010, this was something of a disaster area. Labour was spending £25 billion a year on public-sector IT – more per person than any other country in the world.[116] And 70 per cent of this went to just seven big firms, which is perhaps why the government was being charged £57 for power cables available for £20 on Amazon and £8 wholesale.[117] In terms of the nation's online presence, there were 750 separate government websites, each designed, hosted and created separately.[118] And the internet was simply not on the radar of the people who mattered. At the height of the financial crisis – in the very week that the government was taking the historic decision to bail out the big banks – its main internet portal, direct.gov.uk, was devoting its front page to educating the public on National Bee Week.[119]

Learning from the private sector, and its struggle to innovate within established hierarchies, the Conservatives recruited bright, unconventional hackers to oversee the entirety of the government's online output, and stripped departments of much of their power. The result was the Government Digital Service (GDS), a lean, bare-bones outfit in a set of offices in Holborn that, with its whiteboards and wallcharts, looks as much like a Shoreditch start-up as a department of state.

One of GDS's key insights was that innovation is proceeding so rapidly that the old government model of IT procurement – to sign a very large contract with a very large company and then watch as things go expensively wrong – could simply be bypassed. By bringing the talent in-house, and by constructing simple, scalable prototypes, the team could do in a matter of weeks what had taken years. In the process, they have saved significant amounts of money – by 'taking things that used to cost two billion and delivering them for tens of thousands', in the words of Rohan Silva, the senior Downing Street adviser who spearheaded GDS's creation.[120]

A leading role in this was played by Francis Maude, the minister for the Cabinet Office between 2010 and 2015. When he started work, he recalls, Britain was spending 'vicious' amounts of money on IT – almost all of which came from those seven big firms, via seven-year contracts which took three years even to negotiate.[121] The result was that most 'was obsolete before commissioned, and positively antique by the end of the contract's life'. Payments of agricultural subsidies to farmers, to give just one example, were made using four separate incompatible systems, at a cost of £727 per transaction – and the whole thing was working so badly that Britain had been fined by the European Union.

The agriculture department therefore decided to set out the business case for building a new system – an exercise which was set to cost £6 million before a single line of code had been written. At GDS, meanwhile, a tiny team simply went ahead and built a sample payment app as a proof of concept. The result was a vastly cheaper system. On another occasion, a government contract for IT hosting was for £4 million. A small British business provided a better solution for a mere £60,000.

Maude cites one particular eureka moment – visiting Silicon Valley and meeting a firm of 80 people who had just provided a database for the Indian government containing ID card details for its billion-plus citizens. In Britain, by contrast, those bidding for government work had to provide three years' worth of audited accounts – disqualifying not just start-ups, but anyone imbued with a start-up mentality.

Big government and big companies, notes Silva, 'talk at the same speed', which is perhaps why they feel so comfortable with each other. In David Cameron's first speech on the economy as prime minister, Silva inserted a final section pointing out the scale of churn in the FTSE 100, and how the giant firms (and employers) of the future hadn't even been created yet – hence the need for government to foster small businesses and entrepreneurship. He recalls:

'Whitehall wanted that section taken out. It was such a profound challenge to their world view, which is that big companies are the only ones that matter, big companies are the only ones you should talk to.'[122]

It is important to point out that this digital revolution has not been without its setbacks. In a devastating article for *The Register*, Andrew Orlowski argued that the pruning back and simplification of government websites also resulted in a catastrophic loss of functionality for many of them.[123] Internal memos talk of a breakdown in quality assurance described as 'general chaos' and 'a total nightmare'.

Perhaps the idea behind the team – of creating a crack digital team that could cut through the Whitehall labyrinth – was pushed too hard. But it certainly impressed Barack Obama enough for him to form his own version of GDS, the US Digital Service, personally headhunting many of the staff involved from the Silicon Valley giants. As a test of whether the government was serious about embracing the Silicon Valley ethos, USDS's star recruit Matthew Weaver insisted on the *Star Wars*-referencing job title 'Rogue Leader' – and it was granted without demur.[124]

After the debacle of healthcare.org – when the US government attempted to move the entire population on to a new and untested database without hiring good enough programmers – there is a recognition in Washington and Whitehall that getting better coders will be at the heart of the next generation of public service delivery. For one thing, providing services online saves public money – and voters' time. 'Digital is not just better and more convenient but an order of magnitude cheaper,' says Richard Sargeant, a Google veteran who was GDS's director of performance and delivery.[125]

Chris Yiu of the think-tank Policy Exchange has calculated that in the UK, £24 billion a year could be freed up by getting government to adopt a digital-first mentality.[126] The government's own *Digital Efficiency Report* found that digital transactions are 20 times cheaper than phone calls, 30 times cheaper than sending forms through the post, and 50 times cheaper than a face-to-face talk.[127] In America, a five-man team at the firm Intuit created an app to take a picture of and process your tax return, potentially saving millions of people hundreds of dollars.[128]

Technology can also help improve performance within government. There has been much outrage recently about the government slimming down HM Revenue & Customs – as though, in their determination to give tax evaders an easy ride, the millionaire politicians are laying off the very people meant to hunt them down. But one exasperated minister points out that those being laid off are the people whose form-filling functions are now being handled by computer – and that the government has thereby been able to employ more tax inspectors, not fewer.[129] (What that winnowing out of the professional classes due to automation will do to the tax base is another story, of course.)

Then there is the potential for more imaginative uses of technology. When it came to power, the Coalition in the UK used a combination of online and offline techniques to create the 'Public Spending Challenge' – essentially, to flatten the hierarchies by getting government employees to suggest ways in which the public services could do better with less. More than 110,000 ideas were submitted by civil servants, and 48,000 more by the public.[130] 'The world of policymaking is still fairly limited,' says Mark Flanagan, who helped organise the project. 'You have your Rolodex of the regular vested interests or stakeholders who have access to you. Why not open that up? The technology and tools are available to get a much wider spread of ideas.'[131]

Sadly, the track record of crowd-sourcing policy in the UK and US is mixed. The Public Spending Challenge fell into abeyance, with the idea only recently being revived. Similarly, while both the British and American governments have petition sites, intended to force politicians to debate and consider issues that people genuinely care about, they are far from a perfect tool – the most famous example to date is a petition to persuade the US government to build a copy of the Death Star, which prompted a detailed rebuttal explaining that the cost, even with the Pentagon's defence budget, would be sadly prohibitive.[132]

But other countries are more committed to such thinking. Several towns and cities in South America operate a system called participatory budgeting, whereby everyone gets to vote on spending priorities.[133] An extreme version of this, proposed by the MP Douglas Carswell in his book *iDemocracy* (and experimented with by the Pirate Party in Germany), would see voters able to get involved in politics not just every four years or so, but whenever they felt like it via mass, instant electronic plebiscite.[134]

There is a risk that such a plebiscitary system would embody all the worst tendencies of the great acceleration, and end up mimicking the system that The Circle, the hyper-corporation in Dave Eggers's book of the same name, calls 'Demoxie' – delivering 'change at the speed our hearts demand'.[135] But this could be mitigated by allowing us to delegate our vote on particular issues to a trusted expert or confidant – a friend, a particular campaign group, even a politician. Alternatively, Carswell points out that the US government buys $24,268 of public services for every American – so why not develop hyperpersonalised services whereby we can spend that money as we choose?[136]

There are all sorts of smaller ways that the business of government can be speeded up by the judicious application of technology. In New York, for example, the authorities instituted the '311' helpline for minor civil irritations, such as potholes or parking restrictions. By conducting proper data analysis, as Steven Johnson describes in his book *Future Perfect*, they were able to improve their services: for example by identifying excessive noise as the number-one cause of aggravation, or working out that the first day of spring would bring a surge in demand for CFC recycling, as the warm weather inspired people to upgrade their air-conditioning.[137] The use of crime mapping, similarly, has helped police forces in America and increasingly in Britain to target their resources more effectively, rotating their focus hour by hour.

FUTURE IMPERFECT

In their hearts of hearts, civil servants and politicians nurse all sorts of schemes for pushing the state to embrace an age of agility. 'Maybe you don't have a permanent standing Civil Service,' muses Rohan Silva. 'Maybe you bring in the people you need and get their skills.'[138] One of his former colleagues argues that Britain would be far better ruled with three-quarters of Whitehall given the axe: 'The way to make government more lean, agile, faster is to get rid of vast swathes of departments, vast swathes of ministers.'[139] But, he adds, it will never happen, 'because a) the Civil Service will be institutionally opposed to it and b) for a prime minister, half the number of ministers means half the amount of patronage'. Others talk of eventually handing over management of the economy to a suitably advanced algorithm.

In the more immediate term, however, there is a palpable sense that the current system is breaking down: in Britain and elsewhere, traditional two-party politics is cracking under the strain of a more restless and diverse population. Cynicism towards politicians, and towards a bloated, slow-moving and inefficient government, is rampant. The state has spent too much (and too wastefully) and borrowed too much. The rising powers of the East – especially the Chinese – increasingly view Western democracy as something to avoid rather than envy, given how hard it makes it to actually get things done.

In other chapters of this book, I have tried to show how the problems thrown up by the great acceleration are more than matched by the benefits it offers. But the truth is that when it comes to politics, it is hard to be so sanguine. There is, as Schmidt and Cohen argue, 'a canyon dividing people who understand technology and people charged with addressing the world's toughest geopolitical issues, and no one has built a bridge'.[140] The British writers John Micklethwait and Adrian Wooldridge agree. 'In both Europe and America,' they say in their book *The Fourth Revolution*, 'governments are trying to govern the world of Google and Facebook with a quill pen and an abacus.'[141]

So what will the consequences be? The state cannot simply be swept aside, because there are still millions of people who rely on the services it provides and the security it offers. But it faces ever more severe challenges as the great acceleration kicks into ever higher gear, and the turmoil and uncertainty within and between societies increase.

One near-certainty is that the disconnect between rulers and ruled, and the gap between efficient firms and sluggish state, will continue to grow. This is all the more the case because so many problems, in an interconnected world, require collaboration across borders to resolve – and because, in the face of intractable problems, the temptation to take the easy way out and paper over the cracks becomes almost irresistible. One obvious consequence will be a growth in popular disillusionment, partly fuelled by a hypercritical media – the rise of plague-on-all-your-houses movements, or self-professed revolutionaries disenchanted with the whole political and democratic system.

Yet a less obvious but ultimately more powerful trend will be an increasing divorce between those parts of society that embrace and profit from speed – the highly educated and technologically literate inhabitants of the big cities – and those left behind, forced into low-wage jobs and threatened by automation.

One can see the stirrings of such a movement in the calls for Silicon Valley, or London, to secede from their host nations – for example, in Peter Thiel's vision of 'seasteading', which involves creating floating cities in which an Ayn Rand-style cognitive elite can create its own pure capitalist future.[142] Then there was the speech by Balaji Srinivasan, a biotechnology entrepreneur and Stanford lecturer, in which he suggested that America had become 'the Microsoft of nations', a sluggish former giant crippled by an outdated operating system written 230 years ago.[143]

Even if outright secession is unlikely, the power of these new city-states, peopled by a highly educated, highly articulate and vastly wealthy 'superclass' that can replenish itself with the best, brightest and most driven in society, is set to grow and grow. The result will be a dwindling of the ties that bind rich and poor, north and south, given the prevailing sense among those in the fast lane that their rewards are the direct product of their efforts: a cultural shift from 'No Child Left Behind' to 'Devil Take the Hindmost'. One of the advantages of electronic currencies like Bitcoin, for example, is that it makes it harder for government to find and tax your wealth. That's an excellent libertarian proposition. But who then pays for schools, or hospitals, or disability benefits?

There are no easy answers to these questions. But one solution appears to be to move away from the unitary model of the nation-state, and both to give more power to individuals or communities to race ahead or lag behind as they will, and to find ways for governments to coordinate and cooperate so as to tame sudden crises and smooth out the jagged uncertainties that acceleration brings.

Recently, for example, a group of high-level thinkers set up the Oxford Martin Commission for Future Generations, to report on how international institutions can remain fit for purpose. One of its main suggestions was building in sunset clauses, so that when organisations have become too swollen and bureaucratic, they are dissolved.[144] It is not hard to see how that principle could be applied to all legislation, to ensure that our laws and institutions are flexible and agile enough to cope with a changing world.

Another of its proposed solutions – with particular relevance to cross-border diplomacy, which makes even the most ossified government department look positively sprightly – is to abandon the need for consensus on issues, and allow coalitions of the willing to move forward on environmental, financial or trade regulation. These coalitions could include the mighty new city-states as well as traditional nations – a phenomenon which is already happening as London and New York and Paris swap ideas and sign agreements even as they compete.

Such arrangements do, admittedly, create a risk of free-riders emerging, in the shape of nations or groups who benefit from environmental protections for which others bear the costs, or position themselves to undercut particular standards. But they also diminish the likelihood, which is presently all too real, that a problem such as the degradation of the oceans' fish stocks will have reached the point of no return before ministers have even settled on a discussion agenda.

There are plenty of other ideas. Andrew Haldane of the Bank of England suggests that you might place the watchdog role – on big, long-term issues such as the state of the environment or the stability of the financial system – in the hands of independent, arm's-length institutions that are less vulnerable to political and media whim, although that raises its own questions about democratic accountability and legitimacy.[145] Ian Goldin, director of the Oxford Martin School, argues that government should devolve powers to the lowest possible level, so that local officials can respond more instantly to public demands, freeing up Whitehall to focus on the very biggest issues.[146]

The need for change of some kind is clear – not least since, even if our governments are unable to harness the forces of acceleration, their enemies certainly will. In the future, Schmidt and Cohen warn, 'all governments will feel as if they're fighting a losing battle against an endlessly replicating and changing internet'.[147] They describe how Hizb ut-Tahrir opened recruiting stations directly outside Motorola offices in Pakistan, seeking to recruit tech-savvy jihadists.[148] In Russia, as Evgeny Morozov outlines in *The Net Delusion*, his chilling puncturing of techno-utopianism, anti-immigration groups mashed up ethnic data from the census with a map in Volgograd in order to better coordinate pogroms.[149] Iran, he says, could use the same face-matching technology as Facebook to identify protesters, or the voice-recognition software pioneered for virtual assistants like Apple's Siri or Microsoft's Cortana to distinguish the voices chanting anti-government slogans.[150]

Then there is the danger of computer hacking – as when China recently obtained the vetting histories (which included exhaustive lists of embarrassing or potentially compromising personal information) of pretty much every US government employee in a senior or sensitive position.[151] A lengthy *New Yorker* investigation of cyber-warfare and the West's ability to defend against it concluded with an anonymous quote from one of America's most senior officials, cheerily predicting: 'We're completely fucked.'[152]

The great acceleration is making the world faster, but also more turbulent. Without a proper understanding of this new landscape, or the capacity to respond to it, governments are finding it harder and harder to cope. And there is one final irony. The sole area in which politicians really have grasped the potential of technology is in campaigning. More powerful computers have provided political parties with the tools to analyse voters and their preferences more effectively than ever, and to tailor their messages to them – to the point where the technologist Jaron Lanier has warned that political campaigns may become more like the battles between technology firms, won by the side with the biggest databases and cleverest analysts rather than the one with the most compelling or sensible ideas.[153] That, in turn, will alienate those who are not the subject of those particular messages – or lead to a political debate even more dependent on simplistic slogans and *ad hominem* attacks. It will also heighten the disillusionment felt when the glossy, populist campaigner faces the ugly compromises of government – and the decrepit nature of the government machine.

There are, within the halls of power and outside them, many clever and dedicated people who are trying to address these challenges – but they are fighting an uphill battle, not least because many of these changes will require a process of convulsive upheaval. Yet if we do not try to salvage the situation, we will face a dilemma in which politicians get better and better at campaigning and worse and worse at governing – until eventually the public loses faith in the entire system.

7
Time Is Money

'It is not that humans have become any more greedy than in generations past. It is that the avenues to express greed have grown so enormously.'

– Alan Greenspan, testifying before the US Senate in 2002[1]

Before the financial crisis, there was a general feeling that the financial markets could – and should – be left to get on with things. It didn't really matter how they worked, only that they seemed to have a near-miraculous ability to generate wealth and prosperity.

Since then, we have all learned that the markets are something we should pay very close attention to indeed. Yet remarkably, we hardly understand anything about how they actually operate. In particular, we still have the notion that they are run by people – men in striped shirts and braces, standing in vast trading rooms waving bits of paper, or sitting in well-appointed offices hatching elaborate plans to corner the market in pork bellies or frozen concentrated orange juice.

The truth is, this image is laughably outdated – for there is no area of life where the effects of acceleration have been felt as forcefully or extensively as the financial markets. Today's stock market is a string of ones and zeros, a never-ending tumult of bids and offers, housed not on stock-exchange floors but in anonymous warehouses in Slough, Basildon or New Jersey, each of them containing a stack of computer servers the size of a decent house. And the buyers and sellers are not human beings, but computer algorithms, playing war games against each other at speeds measured in milli- or even microseconds.

In theory, the fact that money is moving around faster is a very good thing. The main justification for market capitalism is that it is the best way we have discovered of making us all richer. This is because it is extremely efficient at shifting money to where it can most effectively be used: cycling cash from shopper to shopkeeper, pension fund to start-up. It is this feedback loop of saving and investment, earnings and consumption, that generates wealth, growth and jobs. And the acceleration of the financial world has – in general – made this process far more effective. Capital flows from place to place, nation to nation, at speeds and volumes that can, as the novelist John Lanchester says, 'induce a kind of vertigo'.[2]

But it has also had two very important side effects. The first is that a smoother, more frictionless, more densely interconnected system is also one in which there is far more potential for sudden and unpredicted crashes. The economic commentator John Kay compares it to tailgating on motorways: all the cars have started driving ever closer together down the road, with the drivers inside congratulating themselves on their skill at the wheel and faster journey times. But when a pile-up happens, the consequences are horrendous, as car after car slams into the next.[3]

These crashes, moreover, are not just confined to the markets, but spill over – as we know all too well – into the wider economy. That's because the systems the banks and hedge funds use for their ultra-high-speed stock-market battles are the same that we rely on for the efficient allocation of long-term capital. It's as if, says the statistician Nate Silver, a city 'decided to hold a Formula 1 race, but by some bureaucratic oversight forgot to close one lane to commuter traffic'.[4]

The second great side effect of the acceleration of finance is that long-term thinking has increasingly been driven out by short-term. Just as journalists demand that politicians make instant decisions, so investors demand instant profits from chief executives. The problem is that what is rational for a trader or investor seeking immediate gain may be irrational for the economy as whole. In this chapter, we will examine the consequences of this transformation – first, over the medium and long term, and then over the very, very short.

SHORT TERM, SHORT SIGHT

The business world has always been populated by those eager to make a quick buck – hence the tendency towards herd behaviour, which magnifies the booms and intensifies the busts. This, by itself, is not necessarily a bad thing: Alan Greenspan, the former head of the Federal Reserve, has marshalled evidence to show that 'periodic irrational exuberances' are the price we pay for higher growth and standards of living.[5] An economy drained of its animal spirits is also one drained of its power to generate prosperity.

But our increasing devotion to speed – our inability to look beyond the immediate horizon – appears to also mean we are losing our sense of memory, and perspective. Instead of long-term growth, the focus is increasingly on tomorrow's bonus or profit statement.

In his book *The Future*, Al Gore cites a survey of CEOs and CFOs in which they were offered an opportunity to make a good investment, which would mean missing their next earnings report. Eighty per cent said they wouldn't do it: the dictates of 'quarterly capitalism', in which all that matters is meeting or exceeding targets for shareholders, had taken over.[6] Other researchers have shown that investment in R&D and new infrastructure has a remarkable tendency to plummet just as executives' share options vest (because it makes firms more likely to exceed their earnings targets and thus makes those shares more valuable, at the expense of long-term profitability).[7]

This doesn't just apply to financial firms. Academics at the London School of Economics have shown, in a meta-survey of 51 separate studies, that focusing on short-term financial incentives damages the long-term performance of companies of all kinds.[8] This is one reason why Unilever has moved to half-yearly reporting, with its chief executive informing investors that if they aren't there for the long haul, he doesn't want them.[9]

We shouldn't blame the executives alone, however. The investors who hold their shares are also increasingly impatient, pushing them for quick gains to juice up their portfolios rather than taking the long-term view. Andrew Haldane, the chief economist of the Bank of England, has suggested that the cause may partly be neurological: that thanks to the effects of the great acceleration, we are too impatient to invest without seeing immediate rewards.[10] Certainly, there has been a collapse in the length of time for which shares are generally held: between 1940 and 2008, the average duration plummeted from four years to two months.[11] The British government has become so concerned about this short-termism that it commissioned an official review, led by John Kay himself.[12]

There are, of course, long-term thinkers left in the markets. The financial writer Michael Lewis devoted a marvellous book, *The Big Short*, to profiling the collection of cranks and iconoclasts who saw the sub-prime crisis coming, laid the appropriate bets, and made billions.[13] Then there are investors such as Warren Buffett, who buy firms they like, then sit on the investments for years. For those with such patience, the short-termism of the markets is actually a blessing, as it makes it easier to find good stocks that are suffering from temporary lows.

The problem is that this career path is gradually getting harder to follow. In 2013, the British investor Neil Woodford quit Invesco

Perpetual to start his own fund. The news would normally have been confined to the business pages, except for one thing: Woodford was the closest Britain has come to its own Buffett, with £33 billion under management and a record of multiplying his investors' money 23-fold over his quarter-century career.[14]

In the Kay Review, Woodford was held up as the model of the long-term investor, holding shares for 12 or 13 years on average.[15] And when he appeared before Kay's commission, he complained eloquently about the short-term thinking that characterises the rest of the industry, not least in terms of the incentives, such as commissions and bonuses, that act against savers' interests. Short-termism, he lamented, 'exists at almost every link in the chain between saver and company'.[16]

But here's the thing: Neil Woodford was almost fired. At the height of the dotcom boom, with shares soaring into the stratosphere, his focus on low debt and strong earnings, which led him to invest in things like tobacco firms and energy producers, was viewed as laughably old-fashioned. Of course, when the tech bubble imploded, he was vindicated: but it was a damn close-run thing. The same thing happened during the mortgage boom, which he also steered clear of: despite his august status, there were rumours that Invesco Perpetual were tempted to move him on.

In short, it takes more character and discipline – and luck – to be a Neil Woodford than to follow the herd. Indeed, in both the boardroom and the stock market, it is as if – in the words of the sociologist Philip Zimbardo – 'we've decided to crown the winners of marathons after each hundred-yard sprint'.[17] The pressure on today's giant firms – as elsewhere in our accelerated system – is not just for growth, but for ever faster growth. This tendency, in part, lay behind the massive and disastrous expansion of banks such as RBS before the financial crisis: in the world of quarterly capitalism, they needed to buy up short-term revenues (via acquisition) to cover up their past mistakes.

A ruthless focus on the balance sheet is not, by and of itself, a bad thing. At its best, this mentality – the constant pressure to perform – keeps a firm in overdrive, alert to any slackening of the pace. Yet the ultra-lean corporate structures it promotes also introduce our old friend fragility, making the whole system vulnerable if just one thing goes wrong in the supply chain. James Surowiecki of the *New Yorker* has written about the debacle of Boeing's 787 Dreamliner, in which an obsession with low-cost efficiency ended up with the design being outsourced to 50 different partners. This resulted in massive communication problems and incompatibilities, culminating in a series of battery fires that grounded the fleet, which had already been delayed for three years.[18]

For the iconoclastic left-wing economist Ha-Joon Chang, the problem is that 'both corporations and governments have been forced to implement policies that produce quick profits, regardless of their long-term implications . . . easier movement of finance has also resulted in greater financial instability and greater job insecurity.'[19] His conclusion is that 'unless we vastly reduce the speed gap between finance and the real economy, we will not encourage long-term investment and real growth, because productive investments often take a long time to bear fruit'.[20]

Indeed, one of the most notable features of modern corporate governance is a huge rise in dividends: rather than investing to foster future growth, firms are appeasing investors by showering them in cash. As William Lazonick, professor of economics at the University of Massachusetts Lowell, argues, this mentality also discourages firms from investing in their employees' career progression or long-term success – the emphasis is on keeping investors sweet via immediate profit, share buybacks, etc. The result has been the erosion of middle-class job opportunities and a culture of what he calls 'profits without prosperity'.[21]

This phenomenon – the eating of the corporate seed corn – is also a primary culprit in the debate over whether, and how, innovation is being squeezed out of the economy. This is not by itself going to derail the great acceleration, given the commitment to disruption embedded in Silicon Valley that I described in Chapter 1. But there are some very respected commentators who claim that while we may be spinning our wheels faster and faster, we're not actually moving that much more quickly.

The venture capitalist Peter Thiel blames government regulation.[22] Alan Greenspan blames the growth of the welfare state, which he argues has deterred people from amassing the private savings that used to be reinvested in the stock market.[23] The economist Tyler Cowen focuses on the effects of automation, and the fact that much of the low-hanging scientific fruit has already been picked, making new discoveries harder and rarer.[24] Others have suggested that, in a world where new ideas are copied more rapidly than ever, it is simply not worth investing in a new product line if you will not have sufficient time to profit before the copycats pile in.[25]

But Clayton Christensen, the world's leading prophet of disruption and innovation, has another theory. The problem, as he outlined in London recently, is not that innovation has been squeezed out of the economy – the problem is the kind we're doing.[26]

Essentially, he says, a focus on inventing new things ('empowering innovation') has been replaced by a focus on making the same things for a cheaper price ('efficiency innovation'). A strong economy, he argued in a recent speech, works like a 'perpetual motion machine', with capital from efficiency innovations being used to fund new empowering innovations, which produce jobs and growth.

But something, he says, has gone wrong. 'Finance has cut this circle . . . it takes efficiency innovations and reinvests them in efficiency innovations. My sense is that over the last 20 years, the number of empowering innovations we've been generating is about a third of what it was in the Fifties, Sixties and Seventies.'

The culprit, in this view, lies in managerial short-termism – and, in particular, the cult of the spreadsheet. The markets tell executives that their job is to generate the maximum return on capital, quarter on quarter, and the figures tell them exactly how to do that: by cutting costs and boosting short-term profits. As a result, 'the way we calculate success makes it impossible for innovators to invest in the kind of things that create jobs, and so it makes it impossible for recoveries to create jobs . . . the finance mechanism hijacks capital and recycles it into itself.' The result is pristine corporate balance sheets – and, in the long term, economic stagnation.

Christensen is not the only one to sound the alarm. Susan Christopherson of Cornell University refers to the changes in the economy since the 1980s as a product of 'financialisation' – the intrusion of Wall Street priorities and techniques into the wider economy.[27] For global firms such as General Electric, Nestlé or Ford, actually producing cars or chocolate bars is a less profitable use of their capital than financial engineering.

'These global mega-firms are nominally in the business of making things,' she writes. 'But . . . manufacturing is, in fact, a weak competitor within these firms; the timeline for profits is too long compared to what can be earned through financial transactions.'[28] Indeed, research shows that capital and R&D investment in manufacturing has decreased substantially in many Western countries.[29] Such a phenomenon also explains why investors prefer 'frictionless' virtual companies which can achieve take-off – and a monopoly position – far more smoothly and easily than those that produce boring old real-world products.

Does this mean the end of the great acceleration? Not exactly: the forces powering it are too strong for that. But it is certainly one reason why the Silicon Valley giants tend to adopt unusual voting structures which leave control in the hands of those – such as Mark Zuckerberg or Larry Page – who can browbeat or simply ignore the markets. Jeff Bezos, too, has won an astonishing amount of leeway from investors to pursue long-term, often quixotic, projects at the expense of immediate profit. The risk, however, is that such firms may increasingly become the exception, rather than the rule.

MARKET MAYHEM

There is a solid case that the markets are not doing their job in terms of allocating capital most effectively for the long term – or at least are not doing it as effectively as they could. But the same phenomena – increased short-termism, decreased patience – are in equal evidence at the other end of the stock market, in the minute-by-minute, microsecond-by-microsecond trading of shares. It is here that the acceleration of finance has had its most dramatic effects, and its most worrying.

There are, for example, all sorts of explanations for the sub-prime disaster and what followed – deregulation in Washington; the recycling of Asian savings into Western mortgage loans; pure old-fashioned greed. But what marked out the crisis itself was its speed. Bad debts and bad decisions built up over decades tore the system apart in a matter of days. The behind-the-scenes attempts to resolve the crisis entered the realm of the surreal: Hank Paulson, the US Treasury Secretary, repeatedly vomited from the stress and lack of sleep; Timothy Geithner, the head of the Federal Reserve in New York, rushed from meeting to meeting, trying to keep multi-billion-dollar firms from collapsing overnight.[30] (Such was his mania for forcing these stricken giants into bed with each other that he was nicknamed 'eHarmony' by their CEOs.)[31] In Britain, the future of the banking system, and its trillions of pounds in assets, was decided in a panicky all-night negotiating session fuelled by £350 of curry from the chancellor of the exchequer's favourite takeaway.[32]

Such scenes, while disturbing, are par for the course with panics and crashes. But in 2008, thanks to the speed of modern communications, they were happening everywhere at once. 'When Lehman

Brothers went down in the autumn of 2008, what we saw for the first time ever, in financial history, was every country on the planet, almost without exception, falling off the cliff,' says Andrew Haldane, who before becoming its chief economist was the Bank of England's executive director for financial stability (or, in layman's terms, the man charged with ensuring that 2008 would not happen again). 'That was radically different from any other financial crisis we've ever seen. The Great Depression, for example, didn't swamp the whole world – it affected different people and countries differently. The speed at which this transmitted was radically faster than anything we'd previously seen, and the scale of that transmission was radically greater.'[33]

In part, this was due to the increasing pace of the news media. Instant communications feed our desire for instant information, but also prove far more effective at transmitting shocks and volatility. People and companies stopped spending, or even withdrew the cash from their accounts, everywhere from Boston to Beijing.

Yet the problems in the financial sector also built up because of a devotion to speed – in this case, to making as quick a buck as possible. Financial innovations such as credit default swaps allowed banks and other players in the markets to make enormous bets with relatively tiny stakes: by the time the crisis hit, a bank could be lending $10 billion with only $160 million in actual assets.[34] The Bank of England has calculated that between 2000 and 2008, every penny of increased profit in the banking sector came not from greater efficiency, but higher leverage – making bigger and quicker bets, not smarter ones.[35]

The pressure for profit (to satisfy the hungry investors described above) created its own momentum, accelerating the market further. Money flooded into sectors that were plainly unsuitable – in particular the sub-prime mortgage market – because these loans were the raw material needed to keep the financial conveyor belt whirring. And those who pointed out that the emperor had no clothes largely found themselves out of a job, because their quarterly profit figures lagged behind those of their rivals.

Still, the financial crisis is in the past, surely? We are older, wiser and a lot poorer now, and are certainly not about to make the same mistakes again.

The problem is that, while the specific dynamics that brought about the financial crisis may have changed, the fundamental impulses have not. There is still a trend towards more risk, and more speed.

Take the most basic way in which traders make money: spotting an inefficiency in the market – a product which is trading at too low a price, say – and profiting from that until it is corrected. Some people do this on very long timescales: Warren Buffett might believe that a particular stock is undervalued and will rise over a matter of years. But others in the City and on Wall Street do it on a moment-to-moment basis, making very many small bets very quickly indeed.

Such firms are engaged in what might be described as an arms race of innovation. In order to spot an opportunity for a profit – a product which has one price on one exchange and another on another, for instance – you need to be either smarter or quicker than the other participants in the markets. That means either thinking up a better strategy, or engineering a faster algorithm. But others will soon catch on. So to make the same amount of money, you need to either find a new opportunity or make bigger bets.

This was the trap that Long-Term Capital Management (LTCM) fell into (the firm was so called because it locked those whose money it managed into three-year contracts – a fact that shows quite how short-termist the markets have become). The firm was founded by John Meriwether, whose team at Salomon Brothers – made up of the very brightest maths and physics graduates – was once the most awesome engine of profit on Wall Street. But as Michael Lewis has pointed out, when the same people set up LTCM, they had to tell their backers (which included many of the big banks) what their strategies were.[36] As copycats emerged, LTCM was forced to take bigger and riskier positions. When Russia defaulted on its loans in 1998 – in defiance of the firm's predictions – LTCM found itself hugely overexposed: the rest of Wall Street was forced to stump up $3.6 billion to bail it out in order to prevent (or, as it turned out, postpone) a wider collapse.[37]

Yes, LTCM's basic strategy – memorably described as 'picking up nickels in front of a bulldozer'[38] – was stupid: as many others would in the years before the financial crisis, it made bets that appeared risk-free only because they ignored the possibility of surprising shocks like a Russian default (or, in the financial crisis, a fall in US house prices). But the dynamic of market competition also pushed the firm to go far beyond the bounds of fiscal sense.

The next complicating factor is that – as LTCM learned – the speed of modern communications means that news of your bad bets, or even just rumours about them, can spread more quickly than ever. For all the iconic power of those images from the financial crisis of savers queuing outside Northern Rock, what regulators really fear these days is a virtual bank run – a bad headline or bad set of results pushing algorithms to short a bank's stock, and savers (and, in particular, corporate treasurers) to withdraw their deposits at the click of a button.

When the Royal Bank of Scotland fell, says *The Times*'s award-winning City editor, Harry Wilson, 'you never saw a queue on the street – chief financial officers around the world got worried by what they saw on their screens, and pulled their deposits with the click of a button'.[39] As a result of this virtual bank run, 'whole sales teams were switched from sales calls to begging calls – telling their clients that if you put back a few million of your money, you'll get a bottle of champagne'.

As more people bank online, says Wilson, the opportunities for this kind of panic mount. And the more highly leveraged an organisation is, the worse the impact will be, as it will need to dump a far higher proportion of its liabilities to make up for the decline in its core assets.

This isn't even the worst of it for the banks. Today, many of their operations are in the hands of 'rocket scientists' like those at LTCM: the 'quants', as they are known. But their equations and calculations are so far beyond the layman that it is almost impossible for their bosses, let alone their regulators or the general public, to work out what they are doing. And as the *Financial Times* columnist Gillian Tett has pointed out, these young, planet-brained geeks are generally unlikely to be good people-managers, or to play politics.[40] You therefore end up with a situation in which the executives who have risen through the ranks have only the flimsiest idea what their subordinates are up to. 'Basically, banks have become technology companies,' says Wilson. 'But for some reason they always promote bankers to the top of them.'[41]

For such managers, it is harder and harder to rein their employees in, no matter how dangerous their strategies may be to the financial system as a whole. As Lewis has written, 'the Wall Street CEO can't interfere with the new new thing, because the new new thing is the profit centre, and the people who create it are mobile. Anything he does to slow them down increases the risk that his most lucrative employees will quit and join another big firm, or start their own hedge fund. He isn't a boss in the conventional sense. He's a hostage of his cleverest employees.'[42]

In the wake of the financial crisis, efforts have been made to fix such problems – for example by linking bonuses to longer-term performance rather than short-term profits. But the burden of risk is still grossly tilted towards employer rather than employee. If you are a young graduate hired by a big bank or hedge fund, the implicit promise is that by the age of 40 or so you will have enough money that you never need to work again.

The logical course, therefore, is always to take the biggest risks available, or at least to double down on the trends of the moment, as the odds that your bets will go bad during your time at the firm are longer than the odds of you driving into the sunset two years later in an Aston Martin. Plus, since there are so many rival firms, you can easily find someone prepared to poach you, or to give you a second chance if you are wiped out.

For these hip young gunslingers, it doesn't matter that it's our pension money they're playing with. Or, indeed, that the long hours they work are probably driving them to make worse and riskier decisions: circling back to the research cited in chapter 2 on the importance of decent sleep and reasonable working hours, it's a bizarre state of affairs that we entrust the most important sector of the economy to a bunch of sleep-deprived adrenalin junkies.

HOT MONEY

One of the ironies of the modern markets is that even developments which are, by and large, a good thing can foster such risk-taking. For example, as technology lowers the barriers to entry to the markets, competition increases. This lowers costs to consumers, and makes the markets more efficient. But it also threatens the incumbents' profits – so they take bigger and bigger bets to keep the cash coming in.

This process applies most glaringly to the big banks. As Wilson argues, there is now almost no part of a bank's activities that someone else is not doing. New entrants such as Wonga, PayPal, Tesco and Sainsbury's are invading the retail side of things, issuing loans to customers, taking deposits or offering mortgages. Big firms, such as Apple, Verizon or General Electric, now finance themselves through issuing their own debt, rather than relying on the banks for lending. And on the trading side, there are hordes of hedge funds or other institutions that can play the markets with equal or greater skill.

Now, this is very much a good thing: breaking the banks' monopoly delivers greater competition, innovation and efficiency.

'It's classic Christensen-style disruptive innovation,' says Haldane. 'It's good for stability, because it makes the system more diverse. It's good for the customer, because it makes the cost of doing business much lower.'[43] Indeed, one of the reasons Europe was much slower to recover from the financial crisis was that its companies were far more reliant on the banks for financing, so were starved of credit, whereas their US counterparts went to the bond markets directly.

Yet it does have adverse consequences. In such an environment, in order to stay profitable, banks must either cut their cost base – by adopting better technology, or sacking people, or both – or increase the size of their bets. One of the causes of the financial crisis was that the regulators who should have been curbing this latter tendency were cripplingly reluctant to interfere with what Alan Greenspan called 'the pollinating bees of Wall Street'.[44]

The financial sector, in short, is built around an appetite for risk. But it is also built around speed – in particular, the sheer frictionless ease with which money flows around the world and trades are made.

Money wants to move quickly. But what's striking is how, until comparatively recently, it wasn't allowed to do so. In 1979, when Margaret Thatcher came to power in Britain, exchange controls limited the flow of money out of the country to £500 per transaction.[45] High interest rates acted as another brake on the flow of money – the lower they are, the faster money flows, because there is less incentive to keep your cash in your bank account earning interest.

The acceleration of financial flows, even over the past few years, has been startling. The 1997 Asian crisis, for example, came about because South-East Asian countries had financed their development via short-term debt: they did not understand, says the economist Jeffrey Sachs, 'how vulnerable they would become to panics and euphoric waves of sentiment, coming from London, New York or other money centres'.[46] Indonesia, Korea, Malaysia, Thailand and the Philippines ended up owing $175 billion in short-term debt – 75 per cent more than their collective dollar assets.[47] When people realised this, there was a stampede for the exit, and an economic collapse.

The same phenomenon of 'superfluid' capital was seen in the financial crisis. When the administrator froze the assets of the British branch of Lehman Brothers, he did so, says former chancellor Alistair Darling, 'because he suspected that $6 billion had been taken out of the London office on the Friday evening and had not come back'.[48] Money now moves so fast that it can literally be in two places at once: some Wall Street firms are able to exploit variations in the precise times at which banks settle their account, sending their cash on an endless round-the-world trip, accruing interest at just the right moment. In some cases, the same money can apparently be earning interest on the same day in two different banks on opposite sides of the world.[49]

What did most to create this world was not the deregulation fostered by Reagan and Thatcher, but the computer. Starting in the 1970s, a series of entrepreneurs spotted that Wall Street was ripe for disruption – full of old-fashioned traders doing old-fashioned deals by old-fashioned pen and paper. When one of these bright sparks, Michael Goodkin, approached the British financier Sir Siegmund Warburg for investment, he was told that the idea of using a computer to manage money on Wall Street was not just plainly preposterous, but utterly so.[50]

But it wasn't. Goodkin set up Arbitrage Management Company, recruiting perhaps the most impressive brains trust ever assembled: three of the initial team, Harry Markowitz, Paul Samuelson and Robert Merton, would go on to win Nobel Prizes in Economics. They had come up with equations that predicted and detected inefficiencies in the markets – but they needed brute computing power to apply the formulae to the real world. And once the initial difficulties were resolved, it turned out to be, as

Goodkin writes in his autobiography, 'like bringing a gun to a knife fight'.[51]

The next step was to use computers not to model the trades, but to make them. Thomas Peterffy is a Hungarian immigrant to the US with a net worth, at a conservative estimate, of $6 billion. That wealth stems from the relentless application of computing power to the financial markets – a story recounted in *Automate This!* by Christopher Steiner.[52]

In 1977, Steiner writes, Peterffy bought a seat on the AMEX exchange for $36,000, giving him the ability to make trades. But the people he employed to make them for him stood out from the crowd: he mostly hired incredibly attractive models, but also Melvin Van Peebles, the writer, director and star of the hit blaxploitation movie *Sweet Sweetback's Baadasssss Song*.

How did these people manage to make trades that beat all of the seasoned veterans around them? Peterffy had produced a cheat sheet: a set of mathematical formulae, painstakingly worked out, which told him how much the things people were buying and selling should actually be worth. Via phones, and later handheld terminals (touch-sensitive devices 30 years ahead of the iPad), he was able to feed those numbers to his traders on the floor – who, by standing out so much from the usual crowd of sweaty middle-aged men, found it much easier to get their orders taken by the brokers.

Soon, Peterffy's company was raking in money hand over fist, expanding across the country. And as technology become more affordable, he was able to get computers to make the calculations that he'd previously had to labour over himself, creating what was arguably the first algorithmic trader. Known as 'the Correlator', this central computer scanned prices across a dozen markets, generating salvos of trades whenever it spotted differences between them.

Peterffy's next big innovation was not just to generate his trades by computer, but to carry them out the same way. In 1987, an officer from the Nasdaq stock exchange visited Peterffy's offices. Given the volume of shares being traded, he expected to see an army of salesmen. Instead, he was led to a single Nasdaq terminal, wired into a hulking IBM computer. 'Where is the rest of your operation?' he demanded. 'This is it,' Peterffy responded. 'It's all right here.'

This was almost certainly the world's first fully automated trading outfit: prices were taken from the wire, and orders spat back out, without any human intervention. To Nasdaq, it was an abomination: their data had been spliced and soldered into a PC motherboard, enabling Peterffy to react far more quickly to market fluctuations than his rivals. He had, as Steiner says, literally hacked the stock exchange.

The infuriated officials told him the whole thing must be dismantled: even if they were generated by computer, all trades still had to be typed in one at a time, via a conventional keyboard. But Peterffy wouldn't be defeated that easily. In a feverish, six-day bout of innovation, he and his team developed a camera that scanned the information on the screen, and sent the information via PC to a nest of rods, pistons and levers hovering above the keyboard – in short, a typing automaton. When the inspector returned, and stood dumbfounded, Peterffy asked if he would feel better if the typing was done by a mannequin, to preserve at least a semblance of human input. The man stalked out, never to return; Peterffy's business racked up $25 million in profit that year, and never looked back.

THE TRIUMPH OF THE NERDS

In the decades that followed, the takeover of Wall Street by eggheads like Peterffy became complete. Soon, the quants had displaced the old-fashioned traders – and by 2011, more than 70 per cent of US stock market trades were generated by the algorithms they created.[53]

To see how universal such strategies have become, I visited Man Group, one of the oldest firms in the City of London. Founded in 1783 as a barrel-making business, it soon acquired a contract to deliver the daily tot of rum to the sailors in the Royal Navy. It later branched out into trading commodities of all kinds, from rum to sugar to cocoa, eventually – over the course of a convoluted corporate history – morphing into an investment-management firm overseeing tens of billions of pounds for its clients.

Some of this money sits in the AHL family of funds, whose trading was overseen, when I visited, by Murray Steel. Steel started out as a classic City trader, in the days when one of the qualifications for the job was sheer physical presence – a friend who applied for a similar job, he recalls, was turned down because he was too small to make himself seen in the trading pits.

At AHL, the investment strategy is based on the principle of trend-following: the mathematically proven observation that stock market prices that are going up will carry on going up, and those that are going down will carry on going down. The catch, however, is that Steel does not make the bets himself – nor do the small army of men sitting in the AHL office, staring intently at their multiple screens. The buy and sell orders flashing across the system – contracts for coffee, ten-year US Treasury bonds, palladium or lean hogs, stretching across 330 separate markets – are all generated by a 'black box', a high-powered computer that studies the market, applying algorithms written by the very clever people in this office, who are hoping their code can identify the right trends fractionally quicker than their rivals'.

AHL's business is not purely about speed: the trading system tends to hold positions for four to six weeks, rather than flipping between them in seconds. Yet its work is a perfect illustration of how computerisation has changed the markets. For instance, the competitive pressure is intense. 'We know that the super-smart algo guys out there could reverse-engineer anything that the market develops within a few weeks,' says Steel. 'So part of the game is to make sure you're always using different brokers, trading at different times, different sizes, to avoid being obvious.'[54] Even though AHL's trades are relatively long-term, it still needs to make them at the best possible price – and to avoid others catching sight of its movements in the markets.

The people in this room are, in a way, prisoners of their own algorithms: as is more and more common in the world of 'big data', they do not know precisely why their ever-evolving system has made a particular trade, only that it has, and that it works. 'The traders will have their own views,' says Steel. 'They'll go: "Why are we buying stocks today? Stocks are overvalued." But we're not paid to have our own views. Back in 2000, we bought gold all the way from 300, where I thought it was a huge sell, up to 1800. I couldn't get it out of my head that I'd always bought gold at between 240 and 260.' Humans, he says, tend to find reasons to justify their losses, rather than cutting them off. Computers are pure and rational: they 'don't react to fear, greed or panic – these are the human emotions that influence how markets behave'.

But what happens when it doesn't work? Having peaked at a share price of 600 pence in mid-2008, after weathering the financial crisis superbly, Man's returns began to fall. There were whispers that it was because AHL's 'black box' had stopped working – that something in the algorithms was broken.

The real explanation, says Steel, was more simple: whether due to quantitative easing or more widespread market volatility, the trends that his virtual trader followed were less in evidence. The result was a flood of money out of Man Group and a collapse in its share price, as fickle investors decided to seek a quicker fix of profit elsewhere, rather than waiting for the system – or the market – to right itself.

This kind of trading frightens some people – the idea that the 'black box' may run amok. But in truth, there is little danger. Robert Harris may have conjured the idea, in his novel *The Fear Index*, of a supercomputer trading on its own (and – spoiler warning – triggering real-world disasters in order to keep up profits).[55] But in reality, if Man Group's algorithms fail, there is no one's money at risk but its investors' – unless the firm happens to be 'too big to fail', which is a whole other class of problem.

TRADING AT THE SPEED OF LIGHT

The real problems with such algorithmic trading come at the other end of the market, with what is known as 'high-frequency trading' (HFT) – using computers to make bets and exploit arbitrage opportunities at the very shortest scale imaginable.

Of course, the ability to get your trades in before your rivals has always been a key driver of profitability on Wall Street. In the early 1990s, for example, Michael Goodkin decided to re-enter the markets after selling his first firm. He recruited some bright Russian physicists, who came up with a better model for derivatives pricing. This was potentially a vast opportunity: derivatives (which are, to a very rough approximation, bets on the performance of a commodity, or share, or market) are traded in vastly higher volumes than stocks and shares. Yet to his surprise, no one was interested. 'All the traders used the same econometric model,' he wrote in his autobiography. 'Making one trader's model more accurate would be like making one clock tell the correct time in a world where everyone else kept the wrong time.'[56] What they wanted was to be able to work out the 'proper' price, under their flawed model, faster than the 16 hours it then took. So Goodkin sold them precisely that: 'the wrong answer, faster'. Within four years, his firm was worth $40 million.

This is the essential truth of the modern markets: time is money. In 2008, Daniel Spivey was asked by a hedge fund to develop an algorithmic trading strategy to exploit price differences between New York and Chicago. He found he couldn't compete, because other companies were hogging the connections between the two cities.

So he built his own. In 2010, his firm Spread Networks shocked Wall Street by unveiling a new link between the exchanges, one that (unlike the existing lines, which followed railway networks) went as straight as possible, digging and blasting and tunnelling its way through any obstacles.[57]

By shaving almost four milliseconds off the time it took a signal to travel between the two, Spread Networks made sure that any trader who wanted to win the speed race had to go through them. It was so successful that some speculated about setting up an even faster route – one that ignored the Earth's curvature, and tunnelled straight through the ground. Although that failed to materialise, another firm did set up microwave towers along the same route, to winnow out an advantage of a few more milliseconds.[58]

This is all part of what Andrew Haldane calls the 'race to zero'.[59] To take advantage of a particular price discrepancy, you have to be the first to spot it and to send your order into the exchanges. That means you need blindingly fast computers, increasingly with the algorithms stamped directly on to custom-made silicon chips to eliminate the need for signals to be sent back and forth to the processor.

And since being first to know the prices, and first to act on them, confers an enormous advantage on any given trader, you have to be as close as possible to the exchanges: the reason why most of the big trading centres in the UK are within the M25, or just outside it, is that the speed of light becomes a crippling disadvantage once you get past a certain radius. You also have to be the fastest at analysing the data you're collecting: hence why some experts believe Wall Street will beat Google in the race to create genuine artificial intelligence, because of the huge financial advantages that would accrue as it outwitted the other players in the market.

For the moment, the highest speeds are to be found by having your computers housed within the exchanges themselves. This is known as co-location, and it is, as Harry Wilson says, effectively authorised insider trading: while the prices are available to all at the same time, those whose servers are sitting next to the exchange's can act on them before the signal has even reached their rivals.

This, of course, has all manner of consequences. For one thing, it means that if you don't have a super-computer of your own, you're going to be beaten to the punch if you try trading. No less an authority than the Archbishop of Canterbury, in his capacity as a member of Parliament's commission on banking standards, has warned: 'The ordinary investor who does not have access to HFT and who is reading the daily paper or following a stock on the web might as well be back in the 17th century in a coffee shop.'[60] Indeed, what HFT does, in a now familiar pattern, is get rid of the middle ground. 'You're either in the high-frequency game, or you're a longer-term investor,' says Wilson. 'If you're a day trader, you're the dumb money now.'[61]

Today, HFT is a huge business. Recently, a Dutch company called Flow Traders floated on the stock market, with a value of €1.5 billion. Its IPO prospectus described how it traded hundreds of billions of dollars a year, across 94 exchanges, sending up to 90 million order messages a day, with all the decisions made by a software system with more than 1.8 million lines of code. But perhaps the most startling statistic of all was this: there had only been a single day in the last two and a half years in which Flow Traders had lost money.[62]

It is this that makes HFT so controversial. To their critics, such firms are not players in the market, but parasites upon it. In Michael Lewis's gripping and hugely influential polemic *Flash Boys*, he accuses them of being the new villains of Wall Street, using a combination of blinding speed and old-fashioned corruption – namely, reconfiguring the workings of the exchanges to their own advantage – to make sure they get a piece of virtually every transaction.[63] By nudging the price of a share up or down when they see a big order coming, they get a tiny cut each time – roughly 0.1 per cent of every transaction.

Lewis's book introduces the reader to HFT by telling the story of Brad Katsuyama, an aw-shucks Canadian he depicts as the last honest man in a rotten town. While working as a trader for the Royal Bank of Canada, Katsuyama finds that he's never quite able to buy stocks at the price he wants: as soon as he places the order, the price pops up.

This turns out to be because he is ordering in sufficient bulk that no one exchange has enough shares available at any given instant: rather than buying, say, 100,000 shares from Goldman Sachs, his automated ordering system is picking up chunks of 10,000, or 1,000, or even 100. But that 100-share offer is bait, dangled by the wily HFT firms. Katsuyama's system will take the offer, because it is the lowest price out there; that tells the HFT firm's system that someone is in the market for such shares; it then races ahead of the original order, buying up the shares Katsuyama wanted and reselling them to him at a profit.

As he traces the origins of this high-frequency culture, Katsuyama (and Lewis) realises the extent to which speed has come to dominate Wall Street. One of his early hires is Ronan Ryan, an expert in maintaining the data centres that house these computerised exchanges. Ryan recounts how, when he was working for a firm called Radianz, one trader called to ask: 'Where am I in the room?' It wasn't enough for this man to be co-located in the data centre in New Jersey – he wanted to be as close as physically possible to the servers. Another trader asks that his cable doesn't wind around the room with everyone else's, but cuts straight across.[64]

Soon an arms race develops, as everyone tries to pickpocket each other: the speed of data switches, for example, falls from 150 microseconds to 1.2 microseconds per trade. In a marketplace governed by speed, the first to seize on any particular opportunity reaps all the rewards – so firms are happy to be milked by the exchanges, for example by being charged $40,000 rather than $25,000 to rent a data pipe that's two microseconds faster.[65]

Like all heroes, Katsuyama triumphs in the end, setting up IEX (which stands for 'Investors' Exchange') to provide honest pricing. It does this by imposing a sufficient delay on orders (350 micro-seconds, achieved by winding 50 miles' worth of fibre-optic cable round and round in a shoebox) that the scalpers and co-locators are defeated. In the process, however, he and his team claim to have uncovered a catalogue of unscrupulous behaviour by the HFT firms. Lewis writes:

> Broadly speaking, it appeared as if there were three activities that led to a vast amount of grotesquely unfair trading. The first they called 'electronic front-running' – seeing an investor trying to do something in one place and racing him to the next ... The second they called 'rebate arbitrage' – using the new complexity to game the seizing of whatever kickbacks the exchange offered without actually providing the liquidity that the kickback was presumably meant to entice. The third and probably by far the most widespread, they called 'slow market arbitrage'. This occurred when a high-frequency trader was able to see the price of a stock change on one exchange, and pick off orders sitting on other exchanges, before the exchanges were able to react ... this happened all day, every day.[66]

These firms, Lewis insists, are not providing valuable liquidity to the markets – which is their theoretical justification. They are interfering in trading that would have happened without them, and capturing part of its value: 'That money is a tax on investment, paid for by the economy.'[67] This argument has won a huge following: the city of Providence, Rhode Island, is suing dozens of Wall Street banks and other financial companies over high-frequency trading activities which it alleges have ripped it off, which could pave the way for countless other lawsuits.

THE CASE FOR THE DEFENCE

The counter-argument offered by the HFT firms about their activities – and, by extension, their role in the market – is simple. They are not trying to rip people off: they are adjusting their prices in response to the latest information. If you are selling shares at 35 pence and notice that they have ticked up to 36 pence elsewhere, indicating that there is demand for them, leaving your own prices at 35 pence is a sure route to bankruptcy.

'If you have the systems capable of taking in massive amounts of data, parsing it, processing it and spitting you out a price to offer for an asset much faster than anybody else, you're a good trader, you're not front-running anybody,' points out Sam Tyfield, an experienced market observer. 'The reason your broker can't execute at a particular price and has to suffer a tick up or down – is that the market's fault or your broker's fault?'[68] Indeed, in most of these transactions, the only people who lose out are other HFT firms, because a huge proportion of the volume of trading consists of robots fighting each other.

Those involved in HFT – an umbrella term which covers a huge spectrum of activities – do not deny that it has its bad apples. But they insist most of the big firms involved are not 'scalpers' but market-makers: their job, often by formal arrangement with exchanges, is to ensure that there are always buyers and sellers for every commodity on that particular exchange at that particular moment. And in order to do this job well, they need to have the most up-to-date information possible.

'I read this book,' says Remco Lenterman, the managing director of the Dutch high-frequency trading firm IMC, 'and I see a bunch of guys who had no clue.'[69] He insists that many of the 'secrets' that Lewis uncovers were, in fact, common knowledge. As for the more fanciful claims of market manipulation, 'These theories come from people who've never actually met a high-frequency trader. But they have these wonderful theories about what we're meant to be doing. Can I guarantee there's no one on the Street who does it? Of course not. But to us, does this sound very logical? No.'

In fact, the picture can be turned on its head, to paint firms such as Lenterman's as victims rather than villains. The IPO accounts filed for leading HFT market-makers such as Virtu show that they make a profit per share of between 0.03 and 0.07 of a cent.[70] Yet exchanges run under an incentive model: they pay market-makers the 'kickbacks' Lewis mentioned in order to make sure that there is always a liquid market (and, of course, to attract more volume and more business). That rebate, as Lenterman points out, is between 0.25 cents and 0.3 cents per share. Yet the market-maker will only keep about a fifth of that as profit – meaning that 80 per cent of the value for the typical deal is captured by the buyer, not the seller.

'If you talk to the largest institutions in the world – to BlackRock, to Vanguard, to State Street – the people who really own most of the securities, who trade billions of dollars a day, they will tell you a very different story,' insists Lenterman. 'In fact, State Street came out with a paper completely disputing Lewis's findings. They say a) the market's not rigged and b) the market's never been cheaper for them to trade in and c) they think electronic market-makers are providing value to their clients.' Far from imposing a tax on the markets, in other words, most HFT firms are helping them to run more smoothly.

There are, however, areas of concern on which both Lewis and his critics can agree. The first is straightforward deception. Many banks operate 'dark pools' – effectively, mini-stock exchanges in which they and their clients can trade in privacy, out of sight of the markets and hence without moving prices. But on finding that they could not compete with the smaller, nimbler high-frequency firms in the open market, many Wall Street banks cut a deal with them: they would allow such traders into their dark pools, in exchange for a slice of the profits.

As Lewis explains, 'a big Wall Street bank really had only one advantage in an ever-faster financial market: first shot at its own customers' stock market trades'.[71] In the open market, the HFT firms would have been competing with each other, and thereby driving overall costs down. But within the dark pools, they were the apex predators. And not only did the banks let the piranhas into their placid little pools, but they tried to hide what was going on. Barclays's LX dark pool, for example, was so profitable it was known internally as 'the Franchise', with bonuses tied to how fast it grew. But it withered on the vine after it was revealed that the bank had been doctoring its PR materials to hide the fact that the largest fish in the pool was Tradebot, an HFT firm.[72]

The next problem is what is known as 'spoofing'. Because the firms' algorithms are all watching each other, ready to leap on anyone else's trade, you can't simply buy or sell the stocks you want – everyone else would leap in and jack the price up. So a technique has developed whereby the trading robots flood the markets with small spoof orders that are often the opposite of the trade they actually want to make – and then, when the price ticks up or down in response, they withdraw those orders and execute the proper trade.

On the day of the 2010 election in the UK, for example, some 19.4 billion shares were theoretically traded. But the number of transactions that were actually completed was ten times lower, because so many bids vanished in a puff of smoke.[73] This not only skews the market, but the huge flow of orders it creates puts a colossal strain on exchanges' systems.

This brings us to the next problem with HFT: the way in which it has distorted exchanges' incentives. In the grand scheme of things, it doesn't actually matter that much that a particular trader has made pennies on a particular deal – especially since, if Katsuyama is right, such scalpers will go out of business as customers flock to his safer alternative. But it does matter when such activity starts to affect the rest of the market.

This phenomenon is, paradoxically, largely a result of an attempt to level the financial playing field. In 2007, the US government passed a law – known as Reg NMS – saying that traders have to execute orders on the exchange offering the best possible price, in order to lower costs for investors. As Lewis says, this meant that many more exchanges sprang up, each offering very slightly different prices – manna from heaven for HFT firms whose *raison d'être* is to capture and exploit such opportunities for arbitrage, and creating a need for market-makers to ensure that each exchange would be able to match buyers to sellers for each and every asset.

This increase in competition meant that the traditional exchanges faced losing much of their market. As their margins and volumes dwindled, they were forced to seek new revenue streams – in particular, high-frequency trading. The Nasdaq exchange in New York went public in 2005; by late 2011, more than two-thirds of its revenues derived, in one way or another, from HFT firms.[74] This incentivised exchanges not only to invent bespoke order types to cater to their new clients, but to focus on the racetrack and to ignore the commuter lane. One of IEX's most important recruits was Zoran Perkov, whose previous job was to keep the Nasdaq running day by day. This meant, as Lewis says, that:

> he was somehow expected to cope with the demands made on Nasdaq's markets by Nasdaq's biggest customers [the high-frequency traders] and, at the same time, keep those markets safe and stable. It was as if a pit crew had been asked to strip down the race car, rip out the seat harnesses, and do whatever else they might to make the car go faster than it ever had before – and at the same time reduce the likelihood that the driver would die.[75]

The outcome was inevitable. On 22 August 2012, Nasdaq suffered a two-hour outage. It was ascribed to a glitch, but IEX believe it was due to the fact that the exchange 'threw vast resources into the cool new technology used by HFT to speed up its trading' – liquid-cooled cabinets, special switches and wires and fibre – 'and little into the basic plumbing of the market used by the ordinary investor'.[76]

The overall evidence about HFT is that, as Lenterman argues, it makes the markets more efficient: at any particular moment, it results in more accurate pricing for particular products, and ensures liquidity so that anyone is able to buy or sell financial instruments without being ripped off (at least, not by more than a few cents). That is not to say that it does not need to be better regulated, to halt abuses. But to focus on the scalpers and spoofers is to miss the real systemic problem with HFT, which is not that it distorts the market, or that it has bad apples in the barrel. It is that when things go wrong, they go wrong very, very fast.

THE FLASH CRASH

The first, and simplest, way that HFT can wreak havoc is when someone screws up. Peterffy ran into this problem when his system started dumping hundreds of thousands of 'put options' without anyone doing anything. His team pulled out all the wires linking their computers to the stock exchange – but the orders kept on coming. Eventually, they realised that every time someone opened or closed the bathroom door in their office, it sent a draft of air rippling across one of their handheld trading devices, triggering the order.[77]

In that instance, Peterffy had time – just – to work out what was going on, and fix it before he bankrupted his firm. His heirs aren't so lucky. Knight Capital, for example, was a giant trading firm, with revenues of $1.4 billion in 2011. On 31 July 2012, it released an algorithm without proper testing; by the following day, a flood of misguided buy and sell orders had sent the stock prices of 148 firms on the New York Stock Exchange (NYSE) into turmoil.

Knight incurred a $440 million loss, lost three-quarters of its value overnight, and was acquired by one of its rivals at a knockdown price.[78]

Similarly, in 2009, a firm called Infinium Capital Management came up with an algorithm that, once it went live, began selling futures as fast as it could, causing the entire market to dip. In 2010, another of Infinium's algorithms, designed to profit from tiny margins on crude oil trades, went equally wild, losing more than $1 million in three seconds. It was reported that, in the rush to beat the competition, the usual two months of testing for any algorithm had been cut down to only two weeks.[79] And in the newsletter of the Chicago Federal Reserve, it was revealed that in 2003 an anonymous US trading firm 'became insolvent within 16 seconds when an employee who had no involvement with algorithms switched one on. It took the company 47 minutes to realise it had gone bust and to call its clearing bank, which was unaware of the situation.'[80]

Awful as such errors are, they do not actually matter from a systemic point of view – just like Man Group's bad year, Knight's collapse is no one's business but its investors'. What really sends shivers down the spine of those involved in the markets is what happens when these lightning-fast algorithms start to interact in unpredictable ways.

The warning signs have been there since 1987, at the dawn of automated trading. To protect their investments, many US firms invested in a primitive form of insurance: computer software that would sell their shares if they started to go down. The problem, as in 2008, was that everyone had bought the same systems. When stocks began to fall, it created a vicious cycle: everyone moved to dump their stocks, which caused prices to plunge, which triggered the software to sell more stocks. The result was known as 'Black Monday'.[81]

Today, that same process can happen at many times the speed – and potentially cause many times the damage. The best-known instance is the so-called 'Flash Crash' of 6 May 2010. At 2.42 p.m., US stock markets began dropping like a stone: within five minutes, the Dow Jones Industrial Average had lost almost $1 trillion in wealth, suffering the largest single-day drop in its history. Shares in some of the world's best-known companies oscillated wildly, gaining or losing billions of dollars in value: Accenture fell from $40 to $0.01; Apple and Sotheby's rose from $250 and $34 a share to $100,000. Within 20 minutes, however, everything was back to normal.[82]

What's most frightening about the Flash Crash isn't that it happened – it's that we still don't know why. The original explanation was that a Kansas City company called Waddell & Reed released an algorithm that tried to sell off $4 billion in futures contracts – 'eMinis' – too quickly; this triggered a domino effect in which other algorithms sent the prices of some associated investments dropping like stones, and chased the prices of others up to stratospheric levels (since they were programmed with basic correlations along the lines of 'If the price of X rises, then the price of Y will fall, so sell Y quickly').

The exact nature of the various interactions, however, is still shrouded in mystery. An investigation was undertaken by US regulators, but even though they took five months to comb through the data, their central findings were discredited within two or three days.[83] In 2015, a 36-year-old Briton called Navinder Singh Sarao – aka the 'Hound of Hounslow' – was extradited to the US on charges of having contributed to the crash by flooding the market with spoof orders. Sarao had apparently made more than £30 million over five years purely from trading on days of high volatility – and he'd built the systems to do it in his parents' home.[84] But no one has yet explained what Sarao did to cause the crash, as opposed to exacerbating it; indeed, the most common reaction to his arrest was one of bafflement.

That does not mean, though, that we are completely ignorant. However it began, one thing that certainly played a part in making the Flash Crash worse was that the institutions meant to provide liquidity – the all-important capital to keep the market flowing – automatically withdrew due to the volatility the crash caused. Liquidity is, in essence, the knowledge that if you are trying to sell something, there will always be a buyer, and vice versa. In an illiquid market, it is impossible to value goods accurately, so prices go crazy.

This is precisely what happened during the Flash Crash: volatility spooked the market-makers into shutting up shop, which bred more volatility as buyers and sellers vanished and prices went mad. The growth of multiple exchanges has only made matters worse: as the NYSE shuddered and jolted under the flood of orders, discrepancies opened up between its prices and others', causing a flood of automated arbitrage attempts.

The lesson of the Flash Crash is that, thanks to its vast scale, vast speed and massive interconnectedness, the market is no longer a market. It is, as a recent report commissioned by the British government put it, 'a complex adaptive ultra-large-scale socio-technical system-of-systems': a vast web so complex – yet occasionally so brittle – that pulling the wrong string can sometimes lead to disaster.[85] And as Andrew Haldane says, 'a feature of these types of web is that ... it's very difficult to make sense of what happens next, because they operate right on the boundary between chaos and non-chaos'.[86]

The idea that the markets aren't just irrational, but unpredictable at the most fundamental level – and capable of tumbling into chaos due solely to the interactions of computer algorithms, rather than in response to real-world depressions and disasters – is terrifying. And what makes matters worse is that with every event like the Flash Crash, which righted itself quickly and with no substantial long-term damage, it becomes easier to brush aside the entire problem.

This, say the scientists who compiled the UK government report, is the same phenomenon seen at NASA before the *Challenger* shuttle disaster: the 'normalisation of deviance'.[87] Something goes wrong that was never meant to – but everything seems to turn out fine. The same thing happens again, and again, and people come to accept it as normal. Then it happens at just the wrong time, and disaster strikes.

For the Flash Crash was not alone. On 30 July 2012, an erroneous order to sell nearly $4.1 billion of a particular commodity – almost exactly the same volume as that eMinis contract – came in three seconds before the market closed.[88] Had it arrived a few minutes earlier, the result could have been even worse than the Flash Crash – for the US government's circuit-breakers (which halt trading when prices soar or plunge too rapidly to be rational) are turned off in the last half-hour of trading. Such an algorithm-driven collapse in US stock prices immediately before the market closed would inevitably trigger similar selling in Tokyo, then again in Europe when those markets opened. The only reason we avoided such a meltdown was pure dumb luck.

The markets, in short, aren't behaving themselves. 'We've had hundreds of mini-Flash Crashes now,' admits Haldane. 'Not as big as the original – but still, hundreds of unexplained oddities in stock prices at very high frequencies. The evidence base already ought to be compelling on that, but there are still those who say "Who cares?" The risk is that you get a maxi-Flash Crash, and I think the fact that people are still in denial about that doesn't help.'

And the faster the markets get, the greater the risks. One of the hottest areas in finance, for example, is hooking your portfolio to the news: extrapolating from the headlines what the impact on stocks will be, and getting in there first. On 6 April 2015, a report appeared on the Dow Jones Newswire that Intel was in talks to buy a chip-maker called Altera. Within a second, an anonymous bot had bought $110,530 worth of options – i.e., making a bet that Altera's stock would rise. By the end of the day, those options had delivered a $2.4 million profit.[89]

But such strategies can backfire, or be manipulated. In April 2013, Syrian government loyalists hacked the Twitter account of the

Associated Press, and sent out a fake report of explosions near the White House. Algorithms wired into the news feed scanned the tweet, predicted that the market would fall, and set about selling $140 billion worth of stocks.[90] Similarly, the reappearance in 2008 of a six-year-old news story about United Airlines going bankrupt briefly wiped 76 per cent off its share price.[91]

Jeffrey Wallis, formerly of Knight Capital, is an expert on high-frequency trading, and has consistently warned of these dangers. 'Are we going to have another Flash Crash event?' he asks. 'I think there's about a 40 per cent chance that it happens. If you go back to where I was five years ago, looking at the technology and the infrastructure, I'd have said about a 10 to 15 per cent chance. I've accelerated my view tremendously from where I was.'[92]

Even if we avoid a full-on disaster, the uncertainty that HFT breeds is having significant knock-on effects. 'Right now I believe there's a confidence crisis in the electronic marketplace,' says Wallis. 'In the US, since the Flash Crash, there's $300 billion of capital that has moved out of the equity marketplace into other asset classes, just because of the fear of instability. That has less to do with corporate earnings and more to do with fear that the market isn't fair and that you can't predict what you're going to get from it.'

MENDING THE MARKET

Is there a way to have the best of both worlds – the speed, fluidity and efficiency of super-fast capital without the risk of sudden, unpredictable crashes? There may well be. For while HFT might be a symptom of the increasing speed of the markets, it is not an inevitable one, at least not in its wilder and riskier manifestations. All kinds of suggestions have been made for how to bring order to the system.

At the most basic level, there are brute-force ideas for putting the brakes on high-frequency trading. The US economist Joseph

Stiglitz has suggested leaving trading positions open for a full second, meaning that being first to the punch would become utterly meaningless.[93] The European Union, meanwhile, has endorsed a financial transactions tax (aka a 'Tobin tax') which would take a bite out of every trade. The justification is partly to make the bankers pay for their sins, by building up an insurance fund against future disaster. But it's also about putting a little grit in the wheels of finance – and dragging the hectic Anglo-Saxon traders back to a more leisurely, Continental pace.

Both of these are, alas, unworkable. Not only would exchanges and bankers howl about the lost profits, but you would have to do it everywhere in the world, or the trade would re-emerge. Capital has a historic tendency to flow to where regulation is loosest: the City of London became an international hub in the Eighties largely because its eurodollar market was a great way to bypass US regulation. And even limited restrictions can easily be worked around: as the technologist Jaron Lanier says, 'to an algorithm, a circuit-breaker or timing limit is just another feature in the environment to be analysed and exploited'.[94]

That does not mean that the idea of speed limits is completely pointless, however. As well as IEX's time-delay strategy, one foreign exchange platform, EBS, has started to introduce programmes that randomise the order of bids received within a given (but still tiny) period. The idea is not to kill HFT, but to discourage 'speed-only' strategies – a tap on the brakes rather than the accelerator, leaving it to the market to decide which approach to support.

Since Black Monday, there has also been an effort to build 'kill switches' into exchanges for when trading gets out of hand. This idea has grown much more popular recently, but according to Jeffrey Wallis, 'it's a bit naïve – once you get to the point where a kill switch is thrown, that means human involvement, but it's moving so fast that a human has no way to react to it. It needs to be much more automated than what we have today.'

Andrew Haldane agrees that 'you probably want someone poised with their hand on the plug as well, but the first line of defence would be something algorithmic'. These counter-algorithms and circuit-breakers would, crucially, apply not once trades were made, but as they went through the system; if any were deemed dangerous, they would simply not go through.

This would demand enormous computing power. But there is talk, among regulators and bankers both, of the equivalent of a Manhattan Project for finance: an attempt to create a shadow model of the markets, in all their complexity, in which algorithms can be fine-tuned and systemic dangers detected. It would be immensely costly – but given the sums at stake should a maxi-Flash Crash occur, there is an extremely strong argument for it.

All this will be pointless, however, without concerted international action. As long as there are multiple exchanges in the world, regulators will need to act near-simultaneously across them, since turbulence in one market can switch to another in a heartbeat. Haldane points out that one of the great flaws of the current European approach to crash prevention is that it does not work on a cross-exchange basis, unlike in the US: 'That's quite important, because otherwise you just end up pushing the stress to another market, and it ends up circling round the system, back to the exchange that's least liquid. If your circuit-breakers are going to make any sense, they have to apply across exchanges as well as within them.'

The irony, of course, is that it was to prevent exactly this that exchanges were deregulated in the first place. 'The rationale for going down the route of decentralised exchanges and high-frequency trading was that this was a way of bolstering liquidity,' Haldane adds.

> That was the totem – that this would make exchanges and market-making more competitive. And on the face of it, in peacetime, liquidity did indeed appear to improve.

> The question is, did that liquidity improve in wartime as well? Or have we created a structure where liquidity was less present when we most needed it – in other words in situations of stress? And my reading of the evidence is that that's exactly what's happened – that what you've created may be a more liquid pool on average, but there's a flood when you don't need it and a drought when you do, which makes for more fragile pricing and greater volatility.

Haldane's solution is to give market-makers like Lenterman enough perks – such as being charged lower fees by the exchanges – that they accept a contractual commitment to remain in the market in the bad times, rather than drawing in their horns. The alternative is to persist with a system that is super-fast, but potentially super-fragile. 'You don't even have to get into "Is this activity socially useful or not?",' says Haldane. 'That's a subjective, almost a moral statement. Just ask "Is this making the structure of finance more or less stable?" And I struggle to see how it's making it more stable.'

As time goes on, and the process of acceleration continues, these problems will only get more urgent. Already, the life cycle of an algorithm is falling from weeks to days, as rival firms reverse-engineer them more and more quickly. They are also, incidentally, spreading beyond the financial markets: on Amazon, the price of an obscure book on genetics briefly reached $23,698,655.93 (plus $3.99 p&p) because two algorithms from two different third-party sellers had been programmed to set a price slightly higher than their rivals'.[95]

In terms of the financial markets, there is a bigger question: what happens when computers are good enough to create algorithms that, in classic Skynet fashion, can design their own successors? It is not hard to see a game of Darwinian evolution being played out in real time, by computer programs that have no conception of the harm they are doing to the markets or financial system. In planning our response, we need to remember a warning from the UK government's report on the complexity of the modern markets: 'For some systems, when they do actually break, they go so catastrophically wrong so superhumanly fast that the safest option for such a system really is to fix it while it ain't broke, because that is the only chance you'll get.'[96]

This is not, it cannot be said strongly enough, an argument that technology and algorithms are in general a bad thing. The danger with HFT, for example, is not that its short-termism will swallow the financial system: studies show there is a definite (and proportionately fairly small) limit on the amount of the value in the stock markets that it can capture. The problem is that, if we do not regulate intelligently, a collapse in that market is likely to spill over into the wider economy.

Yet you do not need to reinvent the stock market – or abolish high-frequency trading – in order to come up with systems that reward long-term thinking, or at least give it a place in the ecology. Jeffrey Wallis's suggestion that we ban stock exchanges from making a profit might be a step too far, but his proposed ban on spoof ordering seems like a way to restore fairness to the markets while stripping out a large part of the froth and volatility.

The situation is similar when it comes to addressing the wider concerns, raised at the start of the chapter, about how the financial markets can deliver capital in such a way as to promote long-term growth rather than short-term profits. Here too, the best solution is not to stamp down on animal spirits, but to adjust the state of the playing field – to incentivise behaviour that accords with the long-term health of the economy, while still leaving room for the market to operate and the great acceleration to proceed.

For example, Britain and America used to have tax and investment regimes that rewarded long-term capital – via lump sums placed in life insurance bonds, which managers could invest on a timescale of decades rather than months. One of Christensen's proposals is to make it much more costly to move capital in and out of companies quickly, and much more lucrative to leave it in.[97] Another is to bring back taper relief, especially for entrepreneurial investments: this saw tax bills dwindle according to the duration for which assets were held.[98] John Kay's report makes a host of other recommendations: companies should stop managing short-term earnings expectations and announcements; mandatory quarterly reporting requirements should be removed; directors' share options and similar incentives should only mature on their retirement from the company, or even later; and the same should be true of asset managers, whose rewards should come in the form of a long-term interest in their fund rather than a short-term performance bonus.[99]

Making such changes will not be easy, or friction-free. The effects of the Great Acceleration on finance and business have already been dramatic – and are set to get even more so. We are living in a world in which a culture of relentless innovation, and the frictionless availability of international capital, will create great wealth and great advances, but also punish mercilessly those people, and countries, who are ill placed to take advantage of it.

But still – if we can make sure that the casino parts of banking are cordoned off, the dangers of Flash Crashes contained, then this vast global web of finance can still realise its potential to do more good than harm. Yes, we're in a race. But it doesn't have to be to zero unless we allow it.

8

Planet Express

'The line between disorder and order lies in logistics.'

– Sun Tzu[1]

For anyone who thrills to the sound of an engine, the Goodwood Festival of Speed is paradise on earth. Each year, hundreds of thousands of petrolheads make the pilgrimage to West Sussex to watch some of the fastest and most beautiful cars ever created twist and turn up the hill climb before being returned to the display area to be admired by the crowds.

For many of those attending, speed is a way of life. On a *Telegraph* assignment, I was able to share a pot of tea with Adrian Newey – chief designer of Red Bull's Formula 1 team, and generally agreed to be to aerodynamics what Mozart was to the piano concerto – before talking to the team behind Bloodhound SSC, an ambitious all-British attempt to shatter the world land speed record (and the 1,000 mph barrier) by strapping the engine from an RAF Eurofighter into a hi-tech carbon-fibre chassis.

Perhaps the most telling moment came, however, when I headed for the exit. As I climbed into my dinky little rented VW Polo, a classic cherry-red Ferrari pulled out in front of me. And there it stayed, for the next 45 minutes and more, as we crawled our way along crowded country lanes. Even when we got on to the motorway, there was still heavy traffic: it was not until we were almost halfway back to London that the magnificent machine could pull away with a resentful roar.

Over the past half-century and more, we have been fed images of speed and liberation – of trains and cars and planes that whisk us from place to place without friction or hassle. Today, instead of jet-packs and rocket cars, a new generation of visionaries promises hypersonic jets that will abolish the distance between continents, or super-fast pneumatic trains that will ferry us from city to city.[2]

Yet for most of us, getting from place to place is no faster, and not much more pleasant, than it was almost a century ago: we sit and fume in traffic jams, endure grinding delays on buses and trains and suffer demeaning security checks at the airport.

The great acceleration is a phenomenon built, ultimately, on movement – on the idea that people and goods, as well as capital and ideas, should be able to flow with quicksilver speed across the globe. But is that really happening? In this chapter, we will explore the one area of our lives where acceleration appears to be stalling – only to see why, when we widen our gaze, the world is actually moving faster than ever.

THE GREAT DECELERATION

In the wake of the Second World War, massive road-building programmes in the West (especially in America) brought our dreams of movement to life. But since around 1980, congestion has been dragging us back down to earth, and wasting billions of hours of our collective lives in the process. In the US, total annual traffic delays rose from 700 million hours in 1982 to 6.9 billion in 2015, at an estimated cost to the economy of more than $160 billion.[3,4] Across the EU, the cost of congestion is estimated at 1 per cent of GDP.[5]

Nor is there much prospect of things improving. The Centre for Economics and Business Research recently predicted that the cost of congestion to the economy would rise by 63 per cent in the UK by 2030, and by 50 per cent in the US.[6] The chief executive of the Highways Agency, which runs Britain's main roads, has admitted that its target speed on busy motorways has now sunk to 40 mph. Things are getting 'slower and slower', he says – and motorists will just have to get used to it.[7] In the US, as Tom Vanderbilt reveals in his fascinating book *Traffic*, becalmed drivers are spending so much time resting their elbows on their lowered windows that skin cancer rates are notably higher on that side of the body.[8]

What's causing all this congestion? Partly, it happens because there are more people and more cars. It's predicted that over the next 20 years, the number of vehicles on the world's roads will double[9] – yet in the West, many of those roads are already being used to carry more than twice the number of vehicles they were designed for.[10] In parts of the developing world, meanwhile, the jams and congestion are on a scale to make a Western driver thank his or her lucky stars. One tailback on China's National Highway 110 lasted for more than ten days, with speeds occasionally as low as 1 kilometre per day.[11]

But, as Vanderbilt says, it's not just that there are more cars. It's that their drivers are taking more trips – especially the parents. In the West, children who used to walk or cycle to school are now being ferried in the car (a legacy of the 'stranger danger' panic of the Eighties and Nineties). Food shopping has moved from local streets to out-of-town mega-marts (although that may now be changing: we are increasingly finding it more convenient to do a few small shops at local stores, or simply to order our food online and have it brought straight to our doors). All told, global passenger mobility – the total length of the journeys we all collectively take – is expected to triple or quadruple between 2000 and 2050.[12]

All this time unwillingly spent on the roads has turned our traffic system into a vast engine of human misery. Research from Britain's Office for National Statistics summarised the consensus: 'Holding all else equal, commuters have lower life satisfaction, a lower sense that their daily activities are worthwhile, lower levels of happiness and higher anxiety.'[13]

There were some variations: the worst effects were felt by those whose journeys lasted between an hour and 90 minutes, and drivers were mildly happier than those taking the bus (perhaps because they felt more in control of the journey). But as a rule of thumb, with every extra minute you commute, your happiness sinks and your anxiety rises. The effects are so strong that they completely outweigh the benefits of having a larger house or higher salary, which are generally how people justify a lengthening of their commute in the first place.[14]

To make matters worse, the malign effects of a longer commute are magnified by our own behaviour. First, as Frank Partnoy points out in his book *Wait*, a longer journey locks us into the workplace: studies show that, rather than leaving earlier to compensate, we actually stay in the office for an extra 35 minutes per hour of travel time, because we can't face the horror of the journey.[15] Also, jams are inherently unpredictable, so we tend to adapt our travel plans to account for the worst-case scenarios in order to make sure we arrive promptly no matter what. Some drivers in America's most crowded cities now have to set aside five and a half times the normal time for a trip if they feel it's important to be there on time, even though that means they will usually arrive ludicrously early.[16]

The implication of these findings is that a rational government would surely focus on a massive programme of transport investment as the most direct way to decrease the collective unhappiness of its citizens and to get society moving again. After all, the daily commute is the single most miserable part of many city-dwellers' lives.

But it's not as simple as that. For one thing, it is an iron rule that more roads create more traffic, as more people buy cars and those with cars already take more trips. For another, if you improve the transport networks in London or New York, say by building more rail routes into the city, you will not see total travel times fall: instead, you will see the rings of gentrification spread out into the suburbs, as people relocate along the new express routes to bigger homes which offer roughly equivalent travel times.

This, as Tom Vanderbilt says, is because human beings appear to have a built-in daily travel-time budget of approximately one hour: ever since we were cavemen, we have constructed and gravitated towards communities in which it takes roughly half an hour to get from centre to periphery.[17] Under economic duress – in the form of congestion and house prices – we are willing to extend that, but save for a few 'extreme commuters', the misery soon becomes too much.

'The golden rule of commuting,' says Lord Adonis, Britain's former transport secretary, 'is that people will do right up to an hour each way. The average time is about an hour in total, but that's the limit of the range – after that it drops off dramatically.'[18] (One interesting side effect of the great acceleration is that there is also a minimum commuting time: as the pressures of work mount, even those fortunate enough to live right by their office will generally build in a trip to Starbucks or similar in order to create a 'mini-commute' of 20 minutes or so, giving themselves time to prepare mentally for the day.)[19]

There are, of course, ways to mitigate the effects of congestion – many of them drawing on the very latest technology. Services that allow you to see when the next bus is coming just by checking an app or sending a text message have saved millions of wasted hours. Meanwhile, in the urban transport command centres, the latest supercomputers are busy tweaking the duration of every red and green light in order to balance competing priorities and maximise traffic flow.

Traditionally, this was done using historical data – general patterns of vehicle behaviour – but the latest systems, such as those installed in Utah, can monitor and respond to traffic in real time, using tools such as metal induction loops buried in the asphalt, security cameras perched on high, or radar sensors alongside the road. US government research found that real-time signals reduced delays by 21 per cent.[20] In a separate study, Stephen Smith, of Carnegie Mellon University, claimed that 'smart' traffic signals that allow each intersection to control its own flow using traffic algorithms, then beam the results to the next junction on the chain, can reduce travel times by 25 per cent and 'idle' time by 40 per cent.[21] The next step will be algorithms that not only react to traffic, but predict it, defusing jams before they even form.

More drastic interventions can also pay off: London's congestion charging scheme cut traffic in the city centre, and inspired an experiment with variable pricing in Stockholm which involved charging drivers small sums ($3 or less) at the busiest times to use major roads that had turned into chokepoints. This led to 20 per cent fewer cars using the system at peak times, and alleviated 100 per cent of congestion.[22]

Set against this, however, are the ever-increasing volumes of traffic – which make such traffic management technologies a case of bandaging the wound rather than healing it. Yet if we look further ahead, there are tentative signs that our love affair with the car is, if not going into reverse, then at the very least stalling.

In surveys of teenagers' favourite brands, cars have started to drop out of the top spots, replaced largely by gadgets or websites: our identity is now bound up with the devices we carry around, rather than the machines we are carried in.[23] Also, with the wider horizons of digital communication, there is less need to travel in person: admittedly, video-conferencing so convincing as to obviate the need for business travel is one of those developments that always seem to be five years away, but we will get there in the end.

The fact that we may be starting to turn away from cars also reflects the wider trend towards greater convenience that we have explored elsewhere in this book. The growth of car-sharing services such as Zipcar, for example, offered drivers an irresistible bargain: to trade the hassle and expense of maintaining their own car for a five-minute walk to the designated parking spot. According to one study, 80 per cent of members of such car clubs sell their own car after joining – meaning that each Zipcar or its like removes 15 private vehicles from the road.[24]

Even this model, however, has been disrupted by one that removes the inconvenience of having to seek out a vehicle. Services such as Uber – flush with billions in venture capital – will bring the taxi of your choice to you at the moment of your choice, for surprisingly small sums. Particularly for those living in cities, this service is immediately enticing, if deeply aggravating to the incumbent cab companies. And it is powered by nothing more hi-tech than a smartphone, affixed to the driver's dashboard and running Uber's proprietary software.

The next big gain in terms of efficiency will come when we do away with drivers altogether. In vast sheds in Google and Apple's Californian campuses, in the halls of Oxford University's robotics department, and in the R&D facilities of the giant car firms, enormous strides are being made in developing automated vehicles.

Their development, indeed, is a case study in the virtues of acceleration, and of the Silicon Valley model of innovation. As Burkhard Bilger showed in a piece in the *New Yorker*, the decision by the Defense Advanced Research Projects Agency (DARPA), the Pentagon's in-house think-tank, to challenge all comers to develop self-driving vehicles resulted not just in a range of entrants from Stanford professors to a group of insurance company employees from Louisiana, but produced more progress within a single year, 2004–5, than the military-industrial complex had managed in 20.[25] Since then, huge strides have been made, not least by using ever more powerful processors to collect and analyse the thousands if not millions of pieces of data collected by these cars' various sensors every second.

At the time of writing, self-driving cars are at what one Google employee calls the 'dog-food' stage: pretty good, but not quite ready for human consumption.[26] That, however, will change rapidly. And we have not yet realised quite how amazingly beneficial that will be.

The most obvious consequence will be a drastic curtailment in congestion. This is because it is not just about the volume of cars: everyone who has ever driven down a motorway is familiar, if unwittingly, with wave theory, the way that a single tap of the brakes by a driver far ahead will ripple back down the carriageway. Self-driving cars should not only be able to avoid such problems, but to bunch up much closer together, especially if they are networked to the vehicles in front of them in such a way as to know when they are about to change speed.

The next great benefit will be a collapse in the number of traffic accidents and deaths. As Bilger says, 'of the 10 million accidents that Americans are in every year, 9.5 million are their own damn fault'.[27] In Britain, such accidents are the leading cause of death among young people, with human error a factor in 95 per cent of cases;[28] worldwide, they kill 1.24 million people a year and injure 50 million.[29] Indeed, one of the leading researchers in the field, Google engineer Anthony Levandowski, has a peculiarly personal reason for wanting to perfect the product: his unborn baby was almost killed after another vehicle slammed into the car his pregnant fiancée was driving.[30]

Yet a decline in congestion and traffic deaths is only the start of the benefits such vehicles would bring. At a stroke, most of the misery and frustration of commuting would be swept away. Instead of staring grimly at the car ahead, we could hold a business conference as we made our way into work, or stick on a two-hour film from our Netflix queue to pass the time on a family trip to the country.

The next question is whether we would actually need a car of our own. Many transport experts see the future of driving, at least in our cities, as a fusion of Google and Uber: self-driving cars that appear when we want them and trundle off when we are done. Some of us might miss having a personal connection with a vehicle of our own – and there would of course still be a market for luxury or vintage cars – but for most of us, it would be enough to load up our individual 'profile', with pre-programmed seat angles, radio stations or even cabin colours, whenever we stepped into our ride.

The gains for the planet, and for the urban environment, would be huge. Our cars currently spend an average of 95 per cent of their time sitting parked on the street or in garages.[31] If we could keep them in circulation, we would in theory be able to use one self-driving car as efficiently as 20 current vehicles.

Of course, we would never be able to reach a state of perfect efficiency: there would be a need to ensure surplus capacity at rush hour, or on those weekends when everyone heads out of town. But even factoring in a certain level of oversupply, we would need drastically fewer vehicles – which in turn means that swathes of our cities could be turned from parking lots to parkland, or new housing, while streets would grow more pleasant and spacious without all the vehicles lined up on their sides.

It's not just the number of cars that would fall, but their environmental impact – not least because we could order the vehicle appropriate to our trip. As Nick Reed of the UK's Transport Research Laboratory says, you wouldn't just be dialling up a car, but the car that suits your needs at that moment. 'If I'm commuting on my own it might be a small pod; if I'm going on a camping trip, it might be a larger vehicle. And it can have the power source that's appropriate as well – it might be an electric vehicle, or a combustion engine.'[32]

The implications are dizzying. In the 20th century we completely reshaped our urban architecture, and in particular our suburban architecture, to reflect the needs and demands of the car. In the 21st, the self-driving 'pod' could have just as big an impact. Cycling or walking, for example, would become far safer, as cars and lorries became less terrifying and unpredictable companions on the road. Teenagers, pensioners and the disabled would find it far easier to get around. Rather than a world choked with the fumes of billions of personal vehicles, our streets will play host to a delicate ballet of thousands of automated people-carriers, sent swooshing about the streets at the touch of a smartphone.

DELIVERING THE GOODS

This vision of a frictionless, convenient, just-in-time transport network may sound fanciful, but in fact it already exists – at least when it comes to transporting things rather than people. Indeed, this is where the real speed lies in our society – not in the showy cars of Goodwood, or the drivers racing past in the fast lane of the motorway, but in the delivery of what we want, when we want it.

Think of that image of a city of self-driving cars, each proceeding smoothly about its appointed task. That would, if you could zoom out far enough, be rather like what the great global logistics system already looks like: a constant dance of trains, trucks, ships and planes, carrying the goods on which we depend from producer to manufacturer, manufacturer to consumer. We often take this for granted, but it is, without qualification, humanity's most impressive creation. The reasons why were best expressed in Leonard Read's 1958 essay 'I, Pencil'.[33]

Read's central point was that there are approximately a billion and a half of these things on the planet, but not one single person knows how to make one. To produce even such a simple object, he wrote, you would need to know how to cut down the cedar trees for the wood; how to produce the coffee to fuel the loggers as they worked; how to build the roads and railways to take the workers into the forest and take the timber away; how to mill and wax and tint the wood; how to procure and apply the lead and glue and lacquer; how to mine and shape and transport the graphite from Sri Lanka; and so on and so forth. Millions of people, Read concluded, had somehow contributed to the production of each HB – guided not by a central blueprint, but by the awesome power of the invisible hand.

More recently, the *New York Times* columnist Thomas Friedman carried out a similar exercise. In his book *The World Is Flat* he traced the supply chain of every part of the Dell laptop on which he was writing it.[34] In the process, he showed how the global logistical network is knitting together the economies of Asia and America, collapsing the distance between countries and the economic differences between them.

In large part, this is down to the 'containerisation' revolution pioneered by the US entrepreneur Malcolm McLean in the Fifties. His insight was that almost any kind of product or produce could be packed into metal containers of standard dimensions (his originals were 8 ft by 8 ft by 10 ft) which could be seamlessly transferred by crane from truck to ship or ship to train. Immediately, he slashed the cost of loading a ship from $5.86 per ton to 16 cents.[35]

Today, cargo ships the size of overturned skyscrapers deliver thousands upon thousands of these containers to billion-dollar, computer-controlled super-ports, from where they make their way to superstores and warehouses via a chain of distribution centres. So shockingly efficient is this system that it has driven down transport costs to roughly 1 per cent of most goods' ultimate sale price – meaning, among other things, that Chinese factory workers can compete on a more than equal footing with their American counterparts.[36]

And as with passenger transport, the great acceleration promises further gains. In the future, every single product will be part of a vast smart network, the so-called 'internet of things', with embedded radio-frequency identification (RFID) tags constantly broadcasting its location and status – and will be carried to its customers by self-driving lorries that do not have to stop for sleep or rest breaks. Amazon's warehouses already operate in pretty much this fashion, with robots dancing round the aisles, carrying precisely the right products to fill precisely the right size of box. The humans in this system are, effectively, part of the machine, fitted with tracking devices that monitor their movements (to the point of it being flagged up to their supervisors if they take a toilet break at a restroom too far away from their workstation).[37]

As ever with the great acceleration, this is not just something being imposed on us: this vast logistical network has been built up in response to our tastes and demands. Indeed, many large firms are already placing billion-dollar bets on trying to grab a share of the market in short-range deliveries. The theory is that the growth of an 'on-demand economy', based on our mania for instant gratification, will see us order every kind of product or service for immediate delivery to our location. Rather than going on time-consuming shopping trips, or taking the time to cook ourselves an evening meal, we will summon a van, bike or drone which will drop off whatever or whoever we desire within minutes, whether that be a crispy duck with pancakes and hoi sin sauce or a cleaner to change our bed linen.

Already, there is a mini-boom in such services, as firms such as Uber, Google and Amazon compete with boutique services such as Zipments or Petal by Pedal to disrupt traditional retailers: some new homes in the UK even come with secure 'Amazon lockers' by the front door to store deliveries. It may even be that the giant out-of-town shopping malls run by Tesco or Wal-Mart turn into distribution centres themselves, bringing our weekly shop to us rather than the reverse.

Again, we see the familiar self-reinforcing cycle at work. Consumer tastes accelerate, so companies learn to turn over their wares more rapidly. The classic example is the fashion industry, where customers' tastes are changing faster than the traditional dance of the seasons. The success of brands such as Zara and Yoox shows that the firms succeeding in the 21st century are those quickest at turning ideas (in this instance, the latest high-fashion trends on the catwalk) into products that the rest of us can buy: Zara, in particular, has prospered largely via production systems that turn catwalk innovation into mass production much more rapidly than its rivals.

Such agility of production also means that, rather than devoting massive resources to summer and winter collections, and being forced to slash prices on the duds, firms can turn out multiple new product lines every year, doubling down on the winners and letting the losers die a quick death. At the same time, however, this very logistical capacity means they are able to feed and foster consumers' desire for novelty, further speeding up the product cycle.

What are the long-term consequences of these trends? Their primary effect will be to make the economy, and our lives, faster and more efficient – but also more fragile. The age-old economic principle of comparative advantage will ensure even greater specialisation in particular goods in particular parts of the world economy. But such lean, efficient systems are always vulnerable to being knocked off course by sudden shocks. Indeed, when the distribution of products – or, even more critically, the production of their raw materials or sub-components – is concentrated in a few giant factories or warehouses, then a single disaster can cascade up and down the supply chain, throwing whole industries into chaos.

In 2011, for example, an earthquake in Japan disrupted the entire car industry by knocking out the factory that supplied many of the big firms with a particular kind of microchip. James Surowiecki of the *New Yorker* has pointed out that the rise of Uber and Lyft is taking the slack out of the economy, which is largely a good thing.[38] But as James Gleick wrote in *Faster*, the same processes that make the supply chain tighter and more efficient also make it more prone to disruption.[39]

This faster, more fleeting world will also foster the familiar tendency towards gigantism. As we saw in our study of the culture industry, it takes an awful lot of paddling under the surface to deliver what people want, when they want it. While the rise of 3D printing (another technology that both results from, and fosters, our predilection for instant gratification) may democratise and decentralise the manufacturing process, there will still need to be many things delivered to our doors – and that will increasingly tend to be done by firms with the scope and muscle to master these vast logistical chains, whether by their size or their technological expertise.

For example, a key weapon in Amazon's war against traditional retailers in its first decade of life was that, in many instances, it did not have to pay local sales taxes, because its distribution centres were in another, lower-tax state – or even another country, in Europe. But then, a few years ago, it suddenly began coughing up. This was not, as some fondly believed, a retreat in the face of public criticism. It was, as the technology writer, Farhad Manjoo pointed out, to pave the way for the construction of a network of hyperlocal delivery centres, allowing Amazon to deliver whatever you ordered in the least possible time – ideally, on the same day.[40] Indeed, Amazon has gone from promising two-day delivery to Prime customers in major cities to same-day to next-hour.[41]

How are smaller firms to compete with that? Or with Google's 'Project Wing', which uses automated plane–helicopter hybrids to drop packages down to the ground on tiny tethers? Astro Teller, director of the experimental Google X division and official 'captain of moonshots', says that the goal is to take FedEx's same-day delivery revolution and turbocharge it to offer delivery within a two-minute window.[42] Plans are already afoot – involving Amazon,

Google and NASA among others – to segregate the airspace above our cities, with high-speed inter-city drones on an upper level and local deliveries below.[43] Suddenly, your neighbourhood pizza company's motorbikes look very out of date.

This leads to another problem. This kind of global logistical network, dominated by a few Darwinian super-predators, is extremely efficient. But it is also relatively opaque. No matter how large a company is – even an Amazon or a Tesco – it will not have full control of its own supply chain, producing every last item it sells itself. Instead, it will rely on contractors to fulfil its orders – who, in turn, will rely on subcontractors.

An excellent report by Michael Hobbes in the *Huffington Post* shows the problems this can cause.[44] On 24 November 2012, a fire ripped through a garment warehouse in Dhaka, Bangladesh, killing at least 112 people – many of whom, he claimed, had asked to leave their stations when the alarm went off and been ordered back by their managers because they were on deadline.

In the wreckage were found labels for Western brands including Wal-Mart and Disney. But they had no idea their goods were being made there – indeed, Wal-Mart had inspected the factory more than a year before, found it was unsafe and banned its suppliers from using it. What had happened was that Wal-Mart had placed an order (for shorts) with a firm called Success Apparel. Success subcontracted it to another firm, Simco. Then Simco subcontracted some of the order to the owner of the unsafe factory, apparently without telling either of the two firms up the supply chain.

Such arrangements are the product both of fast finance, and fast fashion. 'These days,' Hobbes writes, 'there's no such thing as cycles, only products. If a shirt is selling well, Wal-Mart orders its suppliers to make more. If headbands inexplicably come into fashion, H&M rushes to make millions of them before they go out again.' Honduran companies making T-shirts for the West used to have two months to prepare orders – now they get one week.

To fulfil these orders, the giant Western firms turn to giant Eastern firms, 'mega-suppliers' such as Success Apparel, Yue Yuen or Li & Fung. These companies, although vast and sophisticated, often own no factories themselves: rather, they parcel orders out among thousands of different suppliers in order to fulfil them as rapidly and cheaply as possible. It's at the bottom of this vast supply chain, out of sight of the consumer, that you get the unsafe factories. The flipside is that consumers' tastes are catered to more rapidly and cheaply than ever before – and that at least some of the money they spend will end up with the workers who made the goods, creating wealth and opportunity even if it is of an uncertain kind.

The opacity of these vast logistical systems also means that they are vulnerable to manipulation, especially in ways that increase the power of the biggest players. As we have already seen, Amazon uses its cloud servers to host other people's websites, including many of the largest media or retail organisations in the world. But it also does the same with logistics. Rather than attempting to compete with Amazon's sophisticated delivery networks, many retailers now piggy-back on its systems by renting space in its warehouses (from where their goods can be made available to Amazon's own customers in double-quick time, further entrenching the Seattle firm's market advantage).[45]

There is nothing to suggest, in this particular instance, that Amazon is anything other than a benevolent despot. But it is not hard to see that, in just the same way it monitors and pressurises third-party sellers in its Marketplace, its control of the supply chain could give it enormous power to detect and exploit consumer trends, or to hinder retailers with whom it was in competition.

We do not have to look far to see examples of precisely how such vertically integrated monopolies can use their position for their own ends. Consider the case of Glencore, the commodities giant. At the time of its IPO in 2011, its 1,637-page prospectus claimed it had control of more than half the global market in zinc and copper, as well as substantial interests in coal, grain, oil, and pretty much every other building block of the world economy.[46]

China's economy, says commodities trader Peter Brandt, 'could not exist without Glencore, because it is dependent on raw material imports, many of which Glencore plays a major role in trading and producing'.[47] Admittedly, the firm's lustre has been tarnished by a mega-merger with mining behemoth Xstrata, at just the point when commodities prices started to fall, but it is still a multi-billion-dollar entity with huge market power.

And how does it use that power? Glencore, which began as a trading firm, is a master of the financial tools we discussed in the previous chapter. In 2010, the harvests started to fail at Glencore's Russian farms due to a heatwave. This enabled it to know before other market players that the price of wheat would soar later that year, and to place huge bets in the futures markets to that effect. Just to make sure, its executives used their political muscle to lobby the Russian government to impose a ban on grain exports – which duly sent the markets into a panicked spike, with prices rising by 60 per cent.[48]

It is unfair, of course, to single out Glencore – the truth is, as Paul McMahon points out in his book *Feeding Frenzy*, that the flood of finance into commodities (between 2005 and 2008, the amount in index funds increased from $46 billion to $250 billion) has led to a growth of speculation on an enormous scale.[49] While this does not dictate the direction of markets, it certainly makes the swings faster and wilder, magnifying them by between 10 and 20 per cent, according to the best estimates.[50] This has huge knock-on effects, not least on the people who have to buy the products that such commodities actually end up in.

And when things go wrong, the effects ripple across societies. That wheat-price spike in 2010, for example, provoked a wave of countries to engage in panicked import and export restrictions on grain. These helped to send the price of bread soaring and provoked, in large part, the agitation and tumult across the Middle East that would come to be known as the Arab Spring.

It would be stretching things to say that Glencore by itself brought down Hosni Mubarak and Colonel Gaddafi – but it certainly gave events a good prod. And as ever, it was the most vulnerable who suffered: in wealthy countries, even a doubling of the wheat price might see the price of bread rise by only 10 per cent.[51] In developing nations, the price shock is transmitted directly to household budgets – often among people who are already spending three-quarters of their income on food.

If the world is becoming more turbulent and chaotic – and it is – and politicians are struggling to cope – and they are – it is because, as we have seen again and again, fragility and efficiency are two sides of the same coin, and we live in a hyperefficient world. Consumer demand, corporate profit-making and financial engineering draw products from the earth and soil and send them zipping across the world at ever greater speed and in ever greater profusion. We barely even notice this process in operation, yet our comfort and convenience depend absolutely on it going right – on Amazon delivering that book we wanted within minutes, on Wal-Mart having just the right products in stock when we wheel our shopping trolley round.

The combined effect of all of these changes is that Planet Earth is being inexorably converted into a single great network, one designed to translate raw materials of whatever kind into the products that will satisfy consumers' whims as quickly and as cheaply as possible.

GOING GLOBAL

It may seem, given the scope of this great planet-wide logistical network, that it has no further room for expansion. On the contrary. Because there is one thing above all that will give it fresh impetus, and accelerate the flow of goods (as well as people and ideas) around the world even more rapidly: the rise of Asia.

The portrait of Wal-Mart's supply chain above, while perhaps unsettling, is at the same time familiar: it depicts a world in which the East makes and the West takes. Yet remember the fact that in the survey of walking speed with which this book began, China's pedestrians had begun to speed up too? That reflects one of the most important transformations in the global economy over the past decade or so. Put simply, the East has learned to stop waiting, and start wanting.

We all know the astonishing scope of China's economic transformation. In 1978 the average Chinese income was $200. By 2014 it was $6,000 – and the country was exporting the same amount every six weeks as it previously did in a year.[52] This transformation has been as rapid as anything in history: a process of explosive growth and change akin, in the words of one investor, to 'having Carnegie, Rockefeller, J. P. Morgan, and Mark Zuckerberg sitting together in the same room'.[53]

But equally astonishing is the change in China's culture. In the old days in China, greed was most definitely not good. The Mao era saw athletic competitions banned and successful sportsmen retroactively accused of 'trophy mania'.[54] But as Evan Osnos of the *New Yorker* argues in his book *Age of Ambition*, the state forged a new bargain with its citizens in the wake of the Tiananmen massacre: obey, and prosper. Suddenly, slogans appeared along the lines of 'Borrow Money to Realise Your Dream' – and the individual, says Osnos, 'became a gale force in political, economic and private life'.[55] He spoke to a Norwegian sociologist, Mette Halskov Hansen, who spent four years in a school in the Chinese countryside, and observed a 2008 pep rally in which the children chanted: 'Ever since God created all things on earth, there has not been one person like me. My eyes and my ears, my brain and my soul, all are exceptional. Nobody speaks or behaves like me, no one before me and no one will after me. I am the biggest miracle of nature!'[56]

The idea that the individual has the power to shape their own fate is a hugely powerful one – especially when that fate involves an escape from poverty and deprivation. This hunger for a better life is, indeed, a major driving force of the great acceleration, and of the world economy. Success is equated to the acquisition of material goods: on the Chinese dating scene, those 'triple withouts' who lack a car, a house and a nest egg have little to no chance. Even marriage has a strong pragmatic and transactional element: one woman said on a dating show that cut a little too close to the bone for the censors: 'I'd rather cry in a BMW than smile on a bicycle.'[57]

And, just as in the West, speed has become associated with personal and national progress. Osnos gives the example of He Zhaofa, a sociologist at Sun Yat-Sen University, who published a manifesto warning that China needed to speed up. Citing the same studies of walking speed as this book, he noted that the Japanese were powering ahead at 1.6 metres per second, and lamented that 'even American women in high heels walk faster than young Chinese men'.[58]

Wasting time is not something that the new China can be accused of. Indeed, one of the justifications it cites for its undemocratic system of government is that its autocrats and technocrats have the power to blast through the obstacles that prevent Western leaders from getting anything done. At an elite leadership academy visited by John Micklethwait and Adrian Wooldridge of *The Economist* there was a general feeling that China had little more to learn from Washington or London: instead, the model to copy from was the enlightened autocracy of Singapore.[59]

Yet for all its leaders' boasts that their system is more efficient and less bloated than the West's – in short, that it gets things done faster – China's leaders know their position is surprisingly fragile. It depends on their keeping their side of the bargain: on providing more and more prosperity, more and more quickly, to the millions beating against the door. When the engine showed signs of sputtering out, as in the 2008 financial crisis, billions of dollars in debt were happily pumped into the system in a short-term attempt to shore things up – which in turn became one of the factors behind the stock market panic of August 2015.[60]

India is obviously a very different culture from China. But it, too, feels like it is engaged in a race for prosperity – a desperate attempt to drag as many of its population as possible towards a First World lifestyle and status in the shortest possible time. And since India opened up to the world, it has similarly become suffused with a new sense that work can deliver outsize rewards. If they need proof, the chai wallahs and washerwomen have only to look at the global rich list, where, as of 2008, four of the eight richest people alive were Indians.[61]

Among India's new global class, as with China's, the conspicuous display of wealth has become the norm. In Mumbai, Mukesh Ambani – scion of the giant conglomerate Reliance – has built the world's most expensive private residence, a 27-storey affair involving three helipads, a six-storey garage and a room that constantly generates artificial snow.[62] Britain may have invented Twenty20 cricket, as a response to attention spans that could apparently not cope with four- or five-day Test matches, but it was India that turned it into a global phenomenon – a plutocratic riot of fireworks, rock music, celebrity endorsements, million-dollar contracts and slogged sixes.

And it is not just the rich, or even the middle classes, who are being inducted into the spirit of acceleration. The business guru C. K. Prahalad preaches the gospel of the fortune at the bottom of the pyramid – the revenue to be reaped by making tiny profits from the billions in India's slums and villages. Giant firms like Unilever have dutifully developed single-serving packages of soaps, shampoos or snacks, partly to make a profit, but also to train this vast new market in the habits of consumerism.[63]

The end result of this process is simple. In China, India, Vietnam, Bangladesh and across Asia – and increasingly across Africa, too – millions upon millions of people are acquiring the habit of aspiration. They may not be quite as spendthrift as their Western equivalents (indeed, the scale of self-denial which many Chinese or Indian parents will go through to send their children to university is humbling). But they are nevertheless being drawn into the vast logistical web that girdles the planet, and providing fresh impetus for it to develop on an ever larger scale.

SLICKER CITIES

And now we come to the final piece of the jigsaw – where these people will be living. For what we are busy building is not a network of countries, but of cities: these are the nodes, the linking points at which the flows of data, people and goods converge.

When you think about it, the progression is obvious. A global logistics system will be far more efficient if it's connecting a thousand cities of 7 million people, rather than a million villages of seven thousand. These are the places where we can gain access to the speed we crave, to hyperlocal delivery and hyperfast broadband and, more importantly, to jobs and opportunity.

Again, the scale of this process is hard to conceive. Today, there are 28 cities with a population of more than 10 million people.[64] By 2030, the UN predicts that there will be 41 – and more than half of them will be in Asia.[65] In India, the pull of ambition and the push of poverty will see the urban population almost double over the next 20 years, with some 240 million people moving from country to city.[66] China recently announced a plan to build another mega-city around Beijing, containing a third as many people as in the entire United States and covering an area the size of Kansas.[67] In just a decade's time, the country will already have 221 cities of more than a million people – by comparison, there are only 35 in the entire EU.[68]

The future of our species, therefore, is urban. As P. D. Smith writes in *City: A Guidebook for the Urban Age*, 'unless there is some unforeseen global catastrophe, the 21st century looks set to experience the greatest flowering of urban civilisation in human history. By 2050, it is estimated that three quarters of the world's inhabitants will be urbanites, some 6.4 billion people.'[69] That is only just short of the current global population, full stop.[70]

This will be a world not just of cities, but of mega-cities. We can already see the bones of the urban superstructures that will result – Rio and São Paulo groping towards each other across the 250 miles that separate them; Mexico City swelling to consume the centre of its nation; China's eastern seaboard becoming one long stream of glittering lights. It is these vast metropolises that the great logistical network described above is being built to serve. It is to their consumers that the products of the world will be ferried. It is there that all those cars will be delivered, and all those traffic jams will form. Which means that perhaps the most important question facing mankind is: what kind of places will they be?

The nightmare scenario – which appears to be happening in many parts of Africa – is that this great process of urbanisation outpaces economic growth, with many of the new arrivals dumped in slums and shanty towns. This future is a world of Kinshasas – a city of 9 or 10 million inhabitants with virtually no sewage systems or middle class, where one in five adults is HIV-positive and three-quarters of the population cannot afford formal healthcare.[71] Indeed, as its population grows, Africa may need to build dozens of new urban hubs, because so many of its existing cities are so broken.[72]

Yet a more optimistic vision, given the rapid economic progress of recent decades, is of a world of Londons or Berlins – great global cities that are a productive and creative babble of languages and cultures, in which members of the new global middle class can work and play and collide (while the new global elite gaze down on them from their skyscrapers . . .). These will in turn be linked to other creative hubs by ever faster broadband, or routine air travel, which is expected to double in volume over the next 20 years.[73] To use the shorthand from earlier in the chapter, these will be places of zippy, self-driving smart cars rather than grinding, agonising congestion.

And fortunately, as so often with the great acceleration, the balance of probabilities favours the more optimistic position. Indeed, there is strong evidence to suggest that this mass urbanisation will be the best thing to happen in human history. True, mega-cities are greedy and wasteful places: energy use in Beijing per capita is three times the Chinese average.[74] But part of that wasted energy comes from the fact that traditional urban cores are surrounded by gas-guzzling suburbs. The new eco-cities will be much smarter, more efficient affairs. In the custom-built Korean city of New Songdo, there are no garbage trucks, because there is no waste: all the water, for example, is reused and recycled.[75]

The more important point is that cities are also faster places in the best sense. We have already seen how the pace of our lives scales up alongside the size of our community – but so, as research by Robert Levine and others has shown, so do levels of innovation, productivity, income, even happiness and well-being.[76]

This process, according to the British scientist Geoffrey West, obeys a very simple law. The increase in social interactions between citizens in a larger community – caused, in part, by the fact that they are moving more quickly and bumping into each other more frequently – results in the size and sophistication of its economy scaling up much faster than its population. Put someone in a city twice as large, and they will magically gain the ability to be 15 per cent more productive. Even better, economies of scale mean they will do so at less of an environmental cost: when a city doubles in size, its use of resources rises by only 85 per cent.[77]

It's not all good news: the fact that we're more social in large numbers also means that we commit more crimes, contract more diseases and so on. But by and large, this process is a very good thing. 'As cities get bigger,' West told the *New York Times*, 'everything starts accelerating. Each individual unit becomes more productive and more innovative. There is no equivalent of this in nature . . . they are where everything new is coming from.'[78] And, as so often, this process feeds on itself. Larger cities require more bodies and more innovation to keep the system running, pulling more and more people into the accelerated lifestyle.

Over the next half-century, the Asian middle class is set to swell from some 600 million people to approximately 3 billion, becoming the main driver of global demand.[79] For the West, that may be an unnerving prospect, as we find ourselves and our tastes edged out of global pre-eminence: a PwC report predicts that Britain and France will both lose their place in the global economic top ten by 2050.[80] But for humanity as a whole, the world will be a more prosperous, dynamic and creative place, as these productive mega-cities absorb and enrich people from the countryside in their hundreds of millions. Indeed, it is this, more than anything, that will not just entrench the great acceleration, but give it fresh momentum. We may well still suffer from unwanted and unintended consequences, not least congestion and pollution, but overall, the outlook for prosperity is good.

There is, however, one further caveat, and it may be the most important of all. An economy and society based on instant gratification has created a vast and complex logistical engine, devoted to the rapid movement of everything from packages to people. As the world economy grows, so does the complexity of this network, with more and more nations and suppliers acting under the guidance of the invisible hand.

This magnificent, whirring engine is, as I argued earlier, humanity's single most impressive achievement. But it is fuelled by the planet's finite resources – and what happens when they run out? We have an insatiable appetite for more stuff, newer stuff, and faster stuff. The question we will address in the next chapter is: can the planet cope?

9

Racing to Destruction?

'Clearly, there's a limit to what the US stomach can hold. But I've been predicting we'd reach that limit for the last ten years, and it never arrives.'

– Paul Aho, poultry industry analyst[1]

When I came up with the phrase 'the great acceleration', it seemed a perfect shorthand for the remorseless speeding up of human life. Yet I later discovered that the term was already being used by geologists – to describe the impact of that same process on the planet we live on.

In the wake of the Second World War, scientists realised that humanity was transforming the world at an ever-increasing pace. Haphazardly and often unintentionally, we shoved aside Mother Nature and installed ourselves in her place. The 'Holocene', the most recent geological age, has given way to the 'Anthropocene' – named for the species that is busy converting every part of the planet into raw ingredients for the vast economic system described in the previous chapter.

As Diane Ackerman says in *The Human Age*, 'we've colonised or left our fingerprints on every inch of the planet, from the ocean sediment to the exosphere'.[2] Or, as the eco-journalist Mark Lynas puts it, 'Somewhere between a quarter and a third of the entire planetary "net primary productivity" (everything produced by plants using the power of the sun) is today devoted to sustaining this one species – us.'[3]

This is the great acceleration in its widest sense: an enormous hastening and widening in scale of humanity's impact on the planet. This process was, as this book has argued, driven by our remorseless and rapacious desire for convenience. The problem was that, with so much energy devoted to satisfying our immediate desires, there was precious little left to worry about the consequences. We developed chlorofluorocarbons to stop our fridges bursting into flames, only to realise decades later that they were stripping away the ozone layer that protected us from the sun's rays. We found millions of years' worth of fossil resources buried in the earth, and burned through them in decades – in the process, releasing greenhouse gases that threaten our future. We took the fossilised carcasses of ancient trilobites, compacted together into limestone in the depths of the ocean, and used them to paint our walls and brush our teeth.[4] In short, we gratified ourselves in the short term at the expense of the long term – and as our demands became more insatiable, and our time horizons shorter, we did so with ever greater relish.

So here is perhaps the key question about the process that this book describes. Will our appetites end up destroying and degrading the world around us to the point of no return? Or will the great acceleration enable us to find solutions to the problems we've created? To find the answer, the best place to start is with the seemingly simple business of keeping us fed.

APPETITE FOR DESTRUCTION

Thanks to the great acceleration, our appetites, and our waistlines, are growing relentlessly: between 1980 and 2000, the average

American male put on almost a stone and a half – and is busily adding another one or two pounds a year.[5] Eran Elinav, of Israel's Weizmann Institute, describes the worldwide surge in diabetes and obesity as 'the worst epidemic in human history': on current trends, 50 per cent of children born in 2000 will end up obese and one in three diabetic.[6]

Partly, this is because we are eating more – especially more meat. The average amount of chicken eaten by the average American, for example, has doubled since 1980.[7] But it is also because the food we eat is changing. Our supermarket trolleys increasingly contain products designed to deliver the maximum hit of pleasure for the minimum effort, and laced with salt, sugar and fat to make them more immediately palatable. The result is french fries designed, as Charles Duhigg writes in *The Power of Habit*, 'to begin disintegrating the moment they hit your tongue, in order to deliver a hit of salt and grease as fast as possible, causing your pleasure centres to light up and your brain to lock in the pattern'.[8]

As part of this process, there has been a move away from cooking with fresh ingredients and towards more convenient ready meals and pre-packaged foods. In the average American household, cooking time per day is down to 30 minutes, a whole hour less than in 1970.[9] (I am indebted for this statistic and others to Paul Roberts, whose book *The End of Food* was a major source for this chapter, along with *Hungry City* by Carolyn Steel: they should be required reading for anyone interested in the modern food economy.) By 2030, that figure is expected to be down to between 5 and 15 minutes.[10] In supermarkets, the biggest growth category has been foods labelled 'to go' – products which require no preparation, and only the most minimal chewing, which can be scarfed down in a car or in the street.[11]

It is not just that, because of our distracted lifestyles, we do not have time to think about the consequences of what we are eating. It is that, in many cases, it is much harder to do. The presence of fat in the bloodstream both stimulates the appetite and reduces our ability to notice the signals that tell us when we're full. Developments in food science have had similar effects: many sugary products have switched from using actual cane sugar to high-fructose corn syrup on cost grounds. But because it is made of fructose rather than glucose, it doesn't trigger the various mechanisms which tell us when we've had enough.[12]

Increased convenience in our food supply has, in other words, led us to lose contact with food in the raw, to the point where most domestic meals do not contain even one item that is either freshly made or cooked from scratch.[13] In 1965, recipes in *Elle* magazine could take it for granted that their readers would be capable of skinning a rabbit; today, those same readers have probably never seen one without a cute little bow around its neck.[14] Famously, the TV chef Jamie Oliver baffled children by confronting them with basic ingredients such as potatoes, leeks and onions.[15]

This transformation in our diet is governed, as elsewhere, by our craving for the novel. Food companies have made it their business – as McDonald's once put it in its annual report – to monitor consumers' changing lifestyles 'and intercept them at every turn'.[16] It continued: 'As we expand customer convenience, we gain market share.' This mentality has spawned an entire industry of food scientists and customer analysts – of scientists who model every last aspect of how we consume our food, of tastemakers who prowl the world trying to work out whether a sudden surge in popularity for bacon flavouring, or green tea, or local food rather than organic, represents a momentary fad or a lasting trend. Frito-Lay – the company behind Walkers and Lay's crisps, among many other brands – spends $30 million a year on its research centre in Dallas, to make sure its products are perfectly snackable; its laboratory includes a $40,000 chewing simulator.[17]

The difficulty for such firms is that, as elsewhere, we crave the familiar alongside the novel. As Peter Leathwood, head of Nestlé's

Food Consumer Interaction Department, told Paul Roberts, 'human beings were designed to be hugely conservative when it comes to food. When you're a hunter-gatherer and something suddenly tastes different, that's a warning.'[18] In other words, food firms must devote an enormous amount of effort to producing products that are just different enough to tempt, while remaining familiar enough to comfort.

So why take the risk? Because of the demands of corporate profit statements. As food has become commodified, and prices have lowered thanks to relentless pressure from supermarkets (a phenomenon we shall explore later in this chapter), the only way to thrive has been to add value to what you produce. The secret is not to sell grain, but to enrich that grain with a particular healthy vitamin, and brand it as such; not to sell bread or jam, but to reinvent them as the Pop-Tart.

As Roberts writes, such new products:

> are so vital to manufacturers (by the late 1990s, a third of all industry revenues came from recently introduced products) and so unpredictable (just one in three introductions survives for three years) that companies must continuously release new products (currently around 1,500 a month) in the hopes that at least some of them will survive. To deliver this prodigious output of novelty, food companies have become perpetual product machines, continually translating consumer desires into successive waves of high-margin products.[19]

Driving this process on is the fact that the food manufacturers, such as Nestlé or Kraft, have often grown so large that just one success will barely register on their bottom lines. 'Food companies are now so big and operate across so many product lines,' says Roberts, 'that they can no longer survive on just one or two killers a year; they need dozens to, as one analyst put it, "drive the needle" on continued growth.'[20] It's the blockbuster model again: either they make huge bets, or they engage in endless mergers in order to make themselves even bigger and drive up revenue – which makes it even more difficult to produce enough innovation to satisfy the markets.

The food companies' continuing profitability depends on their catering relentlessly to our appetite for convenience. To do this, they will resort to every neurological trick in the book – and worry about the impact on our waistlines later. But it is not just about new products. What we often fail to realise is the extent to which every single part of the food chain has been ruthlessly re-engineered in order to meet our needs.

GLOBAL SUMMER TIME

When you walk into a supermarket, you are most often confronted by rack upon rack of verdant produce: because seeing fruit and veg makes us feel happier,[21] and because if we've stocked up on healthy stuff at the start of our shop we're more likely to splurge on junk food later on.[22] Supermarkets will also do their best do ensure that each one of those bins is kept full to bursting. That's because the one thing most liable to drive consumers away from a supermarket is the absence of the familiar goods they want at that particular moment – which is why, to be safe, grocery firms massively over-order to ensure that each particular product is kept topped up.[23]

Similarly, because we know what 'proper' food looks like, the firms insist on exacting standards from their suppliers: in the UK, avocados must come within half an ounce of the target weight, while green beans for the French market must be completely straight and precisely 100 mm long.[24] In both cases, anything that does not fit the bill is discarded – leading to huge amounts of food waste. (By one estimate, 40 per cent of all the food in the US is simply thrown away, at a cost of approximately $175 billion a year.)[25]

Here is where the logistics networks we were talking about earlier come in. It takes a scarcely imaginable amount of effort to ensure that our food is delivered to us when we want it – to guarantee that fish caught by a Chinese boat docking in a Thai port to supply a Dutch company arrives at the processing centre in Britain at precisely the right time to be combined with the right quantities of sauce and mashed potato for a microwavable fish pie.

At firms like Pennine Foods near Sheffield, the nation's ready meals are assembled in food mixers almost 6 metres wide, before being cooked in ovens the size of lock-up garages, then shipped to stores in 'retail-ready packaging' that can be deposited straight on to the shelf, saving the time needed to unpack the boxes.[26] And such suppliers need to be constantly on alert to satisfy our changing whims – if a sudden period of sunny or rainy weather drives shoppers to the supermarkets for particular goods, they have to be there waiting, lest the fickle consumers take their custom elsewhere.

So powerful is our expectation that particular goods will be available that it has effectively eliminated the seasons. In terms of fruit and veg in particular, we have entered what the food writer Joanna Blythman calls 'permanent global summertime', in which the same product is always available, whatever the time of year.[27] To give just one example, to cover for the inconvenient defects of the seasonal cycle, American supermarkets cycle between using similar raspberries from California, Chile and Guatemala as the year goes on, in the hope that consumers will never notice the difference.[28]

Convenience, however, is a double-edged sword. The products that are delivered under this system are very often not those that taste best, but those best suited to the rigours of the global logistical network. As Carolyn Steel points out in *Hungry City*, we eat Golden Delicious apples not because they are golden or delicious, but because they are steady growers, can be picked early, store easily, travel well and grow in both the northern and southern hemispheres, ensuring year-round availability.[29] The ubiquitous Granny Smiths, she says, 'have similar handling and cropping properties that outweigh their cannonball-like hardness and searing acidity'.[30]

The natural endpoint of this process is a series of global monocultures, in which local forms of apple or bean or any other kind of produce wither away in the face of the demand from global agribusiness for a few select varieties. Steel observes that three-quarters of strawberries sold in the UK are of a variety called Elsanta, which has shoved hundreds of other native variants out of the food chain;[31] similarly, more than 90 per cent of US milk comes from a single breed of cattle.[32] Gary Martin, an expert plant biologist, claims that as a result of this inexorable drive towards customer-friendly uniformity, our cultivated plants have lost 75 per cent of their diversity over the past century.[33]

One major problem is that such systems are hugely vulnerable to disruption. Shortly after the Second World War, the spread of 'Panama disease' – a virulent form of banana blight – meant that we had to switch, as a planet, from using the Gros Michel variety to the inferior Dwarf Cavendish breed that you see in supermarkets today.[34] Since these bananas were all cloned from one particular plant, retrieved from a remote forest in India, they are hugely vulnerable to a similar epidemic. No wonder that, in his valedictory speech as US Health Secretary in 2004, Tommy Thompson said: 'I, for the life of me, cannot understand why the terrorists have not attacked our food supply, because it is so easy to do.'[35]

Even outside of the produce sector, the desire for speedy production and easy transportation tends to trump the demands of good taste or prudent variety. My former colleague Andrew Brown – a writer and amateur baker who has been nursing the same sourdough culture in his fridge for more than a decade – reserves a special odium for the Chorleywood bread process, invented by the British Baking Industries Research Association and now responsible for producing roughly 80 per cent of Britain's bread. In Andrew's words:

> What Chorleywood does with diabolical efficiency is remove bread's most precious ingredient: time. Time is the one thing (along

> with good grain, water and salt) without which you cannot make good bread . . . If you do away with this waiting period and compress the period allowed for fermentation – 'quick ripening', it's called – you get flavourless, characterless bread. You're also likely to have to rely on a whole range of additives such as enzymes and emulsifiers to make the end result half-palatable, as well as massive amounts of industrial yeast so the dough rises quickly. Instead of the gentle kneading techniques that traditional bakers use, Chorleywood relies on super-fast, 'high-shear' mixing: the dough may get so hot it needs to be cooled down with ice. It's then allowed to rest for no more than a few minutes, before proofing for about three quarters of an hour.[36]

The Great British Bake-Off, this isn't.

It's not just bread: the demands of speed, food safety and easy transportation mean that with all kinds of food, the flavour has to be removed before being artificially reinserted. Mass-produced cookies, for example, retain as little as 3 per cent of their flavour after baking, so it is sprayed on afterwards.[37] Foods are also spruced up with extra additives – fat, salt and sugar are particularly popular, since we are conditioned to like those flavours. Roberts points out that sweetener is now so common in America, even in canned vegetables, that consumers there regard unsweetened food as unpalatably bland.[38]

PLANET OF PLENTY

As these examples suggest, the aim of the modern global farming system is often not so much convenience for the customer as for the supplier – the translation of the minimum possible inputs (in terms of time, money and nutrients) into the maximum possible profits. The result has been a bit less crunch in our apple, or sour in our dough. But it has also, taken as a whole, brought us enormous benefits.

The most obvious and important is that it has helped us feed the world – on a scale once undreamt of. Between 1950 and 2000, the global population increased by 2.5 times. Yet food production more than tripled.[39] The most modern farmers produce unimaginably more food than their predecessors: a US farmer is now 2,000 times more productive than a sub-Saharan sharecropper, whereas in 1900, US yields were only ten times higher.[40]

Much of this change came about by a simple accident. As Paul Roberts recounts, in the late 1940s anglers in Orangetown, New York, found that particularly fine specimens of fish were emerging downstream of a factory owned by Lederle Laboratories.[41] It turned out that the company had developed a chemical called tetracycline, an antibiotic used to treat internal infections. Fish exposed to the chemical run-off from the plant had to devote less energy to their immune systems, to fight infections, and could spend it on increased growth. When the drug was added to the feed of turkey, cattle, pigs and so on, they all grew to jumbo proportions as well – and so the antibiotic revolution was born.

Today, such discoveries are no longer left to chance. The input/output equation is supreme across the length of the food chain. Animals have been moved from pastures to sheds, where they feed on grain, not grass (and therefore grow bigger). They are dosed with vitamins, antibiotics and whatever else they need, often to make up for the shortcomings caused by this change of diet. Whatever can be tweaked, from genes to living conditions, has been: chickens, for example, are raised under continuous light, because it makes them feed more and grow more. Roberts records how:

> breakthroughs in genetics let commercial breeders like Aviagen and Cobb-Vantress manipulate most of the factors that govern a bird's growth, from the tendency to distribute, or partition, muscle mass into the breast region (critical for optimising white-meat yields) to the efficiency of the digestive tract (which lets the bird convert

grain into muscle faster). The resulting broiler was a walking meat machine: twice as big as its 1975 predecessor, with breast portions weighing more than half a pound each, and the ability to reach this sumo-like stature with freakish speed.[42]

This has inevitably had consequences for humans. Mass use of antibiotics in agriculture has led to resistant superbugs emerging that threaten human health, as well as weakening the beneficial bacteria that live in our guts, with all manner of knock-on consequences for our well-being.[43] But for the animals concerned, the costs have been even more severe. Chickens now put on muscle so fast – they grow four times more quickly than 50 years ago – that the rest of their anatomy can't keep up.[44] Leg and heart disorders are common, as are diseases with names such as 'flip-over syndrome'.[45]

These birds live their lives, in battery farms, in cages no larger than pieces of A4 paper. And their deaths are no more dignified. As soon as the birds approach maturity, and stop translating extra feed into extra growth with sufficient efficiency, they are dispatched. Unfortunately, the frantic twitching of wings when a chicken dies sends a flood of lactic acid through these mega-breasts, which in turn damages the meat. While efforts are being made to overcome this, it is quicker and more convenient to 'fix it in post', by pumping the meat with salt and phosphate to help it retain water, which has the happy side effect of bulking it up and inflating profits.[46]

Although battery chickens are an excellent example, virtually any species that produces meat or milk, eggs or caviar, has become part of this accelerated production cycle, and has been refashioned as a result. Hens normally spend a few weeks a year replacing their feathers, during which time they cannot lay eggs – so in Taiwan and elsewhere, they are 'force-moulted', by being starved for ten days to shock them into accelerated feather loss.[47] In fish farms, salmon can be packed 50,000 to a cage, swimming round and round developing deformities, parasites and injuries as they scrape against the edges of the cage, and each other.[48] Trout, meanwhile, are rendered 'triploid' – a state of being neither male nor female which prevents them putting energy into their reproductive system, saving on food costs and lost meat from gutting and discarding the relevant organs.[49]

For some, the moral costs are too much to bear. 'What has happened to us,' asks Steve Hilton, David Cameron's former chief adviser, in his book *More Human*, 'that we think it's all right to throw live chicks into a mincing machine just because they're male; that piglets' tails are chopped off and their front teeth broken to prevent "stress-induced cannibalism" and chunks of their ears cut out for identification, all without painkillers; that cows are milked to breaking point so they live out just a third of their natural lives?'[50]

Well, at root, what is happening is that our food – and, by extension, the natural world – has become just another commodity, part of a link in the value chain stretching all the way from raw materials to our satisfied stomachs. And it means that it is subject to the same relentless levels of transformation and innovation as everything else around us. It is also subject, therefore, to the same tendency towards efficiency – and gigantism.

What position you take on this depends on your point of view. You can mourn the decline of small, local farming – or marvel at the processes which ensure that billions of people can buy food more cheaply and easily than at any time in history. The big supermarkets, for example, offer their customers undreamt-of variety at rock-bottom prices – Wal-Mart alone is estimated to have slashed US grocery bills by 10 per cent.[51]

Still, just because this process has benefits for the consumer, it does not mean – as with the logistical chains in the previous chapter – that we should blind ourselves to the consequences. For example, in order to keep their prices so low, and still make a profit, the supermarkets put huge pressures on their suppliers – not least by making them responsible for holding inventories (including all those surplus fruit and veg mentioned earlier), thus minimising the storage space needed in store. In order to cope with this pressure, the suppliers themselves must get larger – both to hold their own in negotiations, and to keep up revenue growth – and more efficient.

The result is a situation in which firms like Cargill or Tyson are controlling huge parts of the food chain: 81 per cent of the US beef market is in the hands of four giant processing companies, including the two mentioned above.[52] The global tea trade is controlled by just three firms,[53] while a single company, Fresh Express, produces more than 40 per cent of America's pre-packaged salads.[54] And while the US has only a tenth as many hog farms as 30 years ago, they are, on average, ten times larger.[55]

The market can still function in such circumstances, if the firms concerned act as competitors rather than cartels. But when they do, that same competitive pressure drives them to raise profits by producing more and more, at cheaper and cheaper prices.

As ever, this is fantastic for consumers, but it does have unintended consequences – and not just from the way that 'super-sizing' dumps that surplus on to our waistlines. In his book *Farmageddon*, Philip Lymbery of Compassion in World Farming points out that a chicken produced under this globalised agri-system contains more palate-pleasing fat than its predecessors, and less valuable nutrition.[56] The typical supermarket chicken now has almost 50 per cent more calories per portion than in 1970 – and three times more fat.[57] Professor Michael Crawford of the Institute of Brain Chemistry and Nutrition claims that the resulting business is 'fat production, not meat production'.[58]

There are other drawbacks to this system. The fact that animals don't have to be in pasture any more means that you are free to house your factory farms where regulation is lightest and subsidies are highest – and the pressure for profit means that no one looks too closely, or even realises, when those at the bottom of the supply chain start chopping up horses instead of cattle to put in the lasagne, just as no one knew who'd been sub-sub-subcontracted to make Wal-Mart's shorts.

There is another parallel with those textile firms. When you have a plant capable of processing a million chickens a week, as many of these giant firms do, you need those million chickens or your profitability rapidly slumps: throughout the food industry, margins are so small and production systems so efficient that any form of disruption cascades across the network much more quickly. That, of course, includes disease: the marvels of the modern globalised logistical network mean that pathogens can make their way from a lettuce field in Canada, or a swine farm in Thailand, right on to our plates before anyone has sounded the alarm.

Even our attempts to fix such problems often only make them worse. Paul Roberts describes the 'farcical' efforts in 2004 to contain an outbreak of H7N3 avian flu in British Columbia.[59] The chicken barns were pumped full of carbon monoxide, which failed to kill the birds but helped push virus particles out of the barns. Huge portable stunners were brought in to electrocute infected birds – but they were designed for older, withered specimens, rather than plump and healthy males, so 'their executions generated huge plumes of greasy, virus-laden smoke, feathers and other poultry particles'. It has been estimated that 69 per cent of pork and 92 per cent of poultry sold in US shops is contaminated with *E. coli*, because it is simply unaffordable to keep it out of the food chain.[60]

The next problem is that the food industry, just like everything else, has become enmeshed with the financial markets. Historically, agricultural commodities were disconnected from other sectors, providing a hedge against their volatility. But a study by Princeton University has shown that Western commodity markets have started swinging up and down alongside everything else.[61] We only have to consider the Arab Spring, as discussed in the previous chapter, to see the potential instability that can cause.

FROM FEAST TO FAMINE

The most fundamental accusation levelled at our agricultural system, however, is not that it is cruel, or unfair, but that it is unsustainable. We in the West have cleaned our air, restored our rivers and forests, and established pleasant green spaces and wildlife retreats. But that is partly because the global supply chain allows us to export the environmental damage of our lifestyles – or else hide it from sight. We do not see the dead zones from fertiliser run-off, or the sewage lagoons that are the by-product of factory farming's CAFOs ('concentrated animal feeding operations'), even though a single hog or dairy farm now generates as much excrement as a small city.[62] (In 2007 Duplin County in North Carolina, which raises 2.3 million hogs a year, produced twice as much waste as the whole of New York.)[63]

Even the Green Revolution, the agricultural miracle that enabled us to feed the soaring populations of China and India in the second half of the 20th century, was brought about as much by increased use of fertiliser as by the adoption of improved strains of rice and wheat – and that piling on of fertiliser is increasingly suffering from diminishing returns, as the basic nutrients in the soil itself are slowly leached away. California has a reputation as the fruit basket of the world, but according to Philip Lymbery, this is a 'multi-billion-dollar conjuring trick', using water and chemicals to extract harvests from soil 'so depleted of natural matter it might as well be brown polystyrene'.[64]

The pattern is the same elsewhere. In both the developed and developing worlds, rainwater has proved insufficient for our agricultural needs, so we are draining our aquifers – what one might think of as 'fossil water' – at an ever increasing rate. In some parts of India, water tables are falling by 20 ft a year, to the point where specialist oil-drilling equipment is needed to locate it.[65] Similar techniques are being used in California, where drought also stalks the land – but the more the machines suck up, the less is left for the future.[66] Beijing has already sunk by several metres because so much water has been pumped from under China's great northern plain.[67] The Ogallala reservoir beneath the Great Plains in the US, which provides 30 per cent of the country's drinking water, could well dry up totally within 25 years.[68]

The problems are not just on land. We are trawling the oceans for fish with such vigour that more than half of wild populations are, according to the UN Food and Agriculture Organisation, fully exploited.[69] Much of this catch is made up of small fish turned into meal or oil to feed other fish, or chickens or pigs; this is now running so short that some are suggesting the process should be inverted, with the fish fed on ground-up animals.[70] As an example of robbing Peter to pay Paul, it is hard to beat.

It was, of course, entirely natural to use the world's resources to make our lives more comfortable. And we have always been very good at ignoring long-term costs when presented with short-term gains – a tendency which the great acceleration is only exacerbating. Yet we have only just woken up to the extent to which we are stealing from the past and the future to feed our present appetites. In the short term, this threatens to cause political and economic instability, not least 'water wars': Thomas Friedman, the *New York Times* columnist, has pointed out that the Syrian civil war was preceded by the worst drought in decades, driving thousands off their land and into the seething cities.[71] In the long term, the consequences could be widespread ecological trauma.

So what is the state of play? As of 2008, humanity was consuming 30 per cent more per year than the planet sustainably produces.[72] We fill the gap by digging ever deeper into the resources left by history and geology: for example, by 2050 we will have cleared away 70 per cent of the world's tropical and coniferous forest, loosening and degrading the soil in the process.[73]

Not all of our natural resources are at exhaustion, of course. The exploitation of shale gas and oil has transformed the energy markets, staving off the arrival of 'peak oil' for years, if not decades. But even then, there is a sting in the tail. However plentiful the Earth's fossil fuel resources, we cannot actually use them all – because if we did, we would turn the planet into a sauna, as all but the most extreme climate sceptics acknowledge.

Indeed, even the techniques that we turn to in order to save us often come with unintended consequences. When we realised that air pollution was killing hundreds of thousands of people, we cracked down on aerosols – only to find that the haze they produced had been restraining global warming, which suddenly went into overdrive.[74] And it wasn't even as if we solved the basic problem of air pollution, which was too many people driving cars. Oxford Circus in London is now one of the most polluted places in Europe, thanks to the particulates from thousands and thousands of engines.[75] In China, where the problem is even worse, people now have to go on 'lung washing' holidays to clear the filth from their systems.[76]

The most important problem we face, however, is beyond doubt climate change. Indeed, this issue above all is emblematic of our losing battle to resist temptation – to put the long term above the short. And, like the rest of the changes in our lives, it is also packed with unexpected feedback loops, and side effects that feed off each other in unanticipated ways.

Just as the effects of the great acceleration are not simply about the speed at which we go, climate change is not simply a matter of how much carbon we pump into the atmosphere. There are broad causes and effects: for example, the more carbon we dump into the air, the more is captured by the oceans, and the more acidic they become (their pH is now probably the lowest for 20 million years).[77] But there are also sudden, sharp tipping points.

As the ice melts in Siberia or the Arctic Ocean, the darker surface revealed radiates more heat, accelerating the process – as does the fact that the melted water acts as a lubricant to help more ice slip away. And for all that we worry about melting ice raising sea levels, an equal concern is a thaw in the northern permafrost, which contains 1.5 trillion tonnes of trapped carbon, double the entire CO_2 content of the present atmosphere – or of a similar thaw releasing the ancient methane trapped by the Greenland ice sheets, which is 20 times more potent a greenhouse gas than CO_2.[78]

All of this makes an uncertain and unsettled world suddenly seem all the more unstable. The global plenty of the 20th and early 21st centuries was largely built on three great advantages: cheap energy, cheap water and a pleasant climate. All of these may well come under threat in the coming years. For example, scientists estimate that by 2050 there is an 80 per cent likelihood of a 'megadrought' across the American West and Midwest, assuming that we go on dumping CO_2 into the atmosphere at the same rate.[79]

Even without climate change, experts have suggested that the 20th century was an unusually and anomalously stable era, meteorologically speaking.[80] The smallest swing back towards volatility would spell disaster for the delicate, highly specialised monocrops with which we have covered the planet. And, as ever, any disruption will hit the poorest hardest. Richer nations will be able to care for themselves through sheer weight of wallet, but the grain ships that divert into Chinese or American ports in a time of scarcity would leave African bellies empty.

For many environmentalists, these problems are the inevitable consequence of making the natural world a part of the supply chain: it becomes just another commodity to be used up as rapidly as possible. And looking to the future, it is impossible to see the pressures on the system easing. For one thing, there will be billions more people. There may not be quite as many as was once feared, partly because of the strides made in educating women in developing nations and partly because rich urbanites have fewer children than poor villagers. Yet the very fact that the world will be richer comes with its own set of problems.

As we saw in the previous chapter, increasing numbers of people in Asia and Africa are aspiring to share the living standards of the West. But to feed one US citizen, and absorb the pollution they create, takes an area of the planet's land surface equivalent to 9.5 hectares; the figure for Europe and Japan is between 4 and 5, and the global average is just 1.8.[81] If we were to extend the current American lifestyle worldwide, we would need a planet between two and four times the size, not least to produce enough water and grain to feed the billions of extra cattle, pigs and chickens that would need to be raised and slaughtered.[82]

It may be impossible for us all to live like Americans – but it does not mean that we are not trying our best. McKinsey has estimated that by 2030, global water demand will have swollen from 4.5 trillion to 6.9 trillion metric tonnes.[83] This would exceed current supplies by roughly 40 per cent, necessitating huge projects (of the kind already under way in China) to relocate water from wet areas to dry. By the same date, the world could need half as much food again – not because of extra mouths to feed, but more prosperous lifestyles.[84] That is before you consider the thousands more power stations needed to supply electricity, or all the rest of the infrastructure to support us.

One way to make such a future possible is to use energy more effectively. Over the past century, we have taken huge strides in terms of energy efficiency – yet we have also cashed in such gains in the name of greater convenience. The fuel economy of the average American car, for example, has barely shifted since the days of the Model T Ford: engines are massively more efficient, but they are used to power larger and more luxurious vehicles, packed with all manner of gadgetry.[85] Make it cheaper to heat and power our homes, and we will fill them with more devices and larger screens.[86]

REACHING THE LIMITS?

Some people, surveying this gloomy situation, have come to the conclusion that the only sustainable solution is a vastly smaller human population, to lighten the load on the planet – perhaps as few as 2 billion, if we want to be absolutely sure that we are not taking out more from the earth than we are putting in.[87] That, or somehow imposing restrictions on development that prevent us from leading ecologically damaging lives. Or else embracing a culture of slowness, and retreating from our fast-paced technology to more tranquil and fulfilling lives.

At its most extreme, this manifests in the kind of utopian authoritarianism seen in books like *The Value of Nothing* by Raj Patel, held up by some on the left as a set text for remodelling the world economy. He pronounces: 'If meat's to be eaten at all (and I'm not sure it should be), the global allowance will be 25 kilograms of meat and 50 kilograms of dairy per person per year – any more, and the climate will suffer . . . that means, at most, two sausages, one small chicken piece and a small pork chop a week, and milk for cereal and tea.'[88]

Yet everything we have seen in the previous chapters suggests that getting people to accept such limits will be an uphill challenge to say the least. And even if we in the West can restrain ourselves, will the billions of middle-class consumers in China and India really stop buying cars and meat on our say-so? Who, indeed, will tell the struggling villagers of Africa that they must sacrifice themselves for the global good? Across the world, roughly 1.5 billion people have no access to electricity, and another half-billion make do with coal-powered cooking fires.[89] Can we really deny them access to the fossil-fuel economy, even if it accelerates global warming?

Rather than hoping we can convert human nature into something other than it is, we must work with its demands and limitations. Which brings us back to Reid Hoffman's observation about how to build a start-up – jump off a cliff, and assemble a plane on the way down. But instead of a company, it's the planet that we're risking. And given the futility of tapping on the brakes, our best solution is to invent some wings – fast. Because uncomfortable as it is to admit it, the only thing that can realistically fix the planet's problems, and sustain not just the present population, but the vastly larger and richer population to come, is what landed us in this ecological mess in the first place: technology, innovation and trade.

On the surface, that might sound like a crazy argument, especially if you believe that accelerated capitalism is itself the problem, rather than its unwanted side effects. But there are hugely encouraging signs that we will be able to overcome our problems – and, indeed, that those problems may not be as severe as many people fear.

To return to the topic of food production, for example, the truth is that humanity is already growing enough to feed itself more than adequately.[90] The problems are, first, that much of it is diverted to make biofuels, or fatten up pigs, chickens and cows, and second, that an enormous amount is wasted. In total, 30 per cent of what we produce is simply thrown away, either on the farm, by the supermarkets or by consumers.[91] Like embracing energy efficiency, cutting food waste is a win–win, because it saves money even as it helps the planet.

Nor should we fear running out of food, at least in the immediate term. Currently, around 12 per cent of the planet's surface, or 1.6 billion hectares, is used for agriculture.[92] In his book *Feeding Frenzy* – another invaluable source for this chapter – Paul McMahon cites research by Gunter Fischer of the International Institute for Applied Systems Analysis which estimates that there are another 1.3 billion hectares easily available, and even more if we are willing to encroach on forests or natural parks.[93] As McMahon points out, there are vast 'land banks' in Russia and Eastern Europe, and especially in South America or Africa, both of which could easily triple their amount of land under cultivation.[94]

Then there is the yield gap. At the moment, says McMahon, Western Europe and East Asia are using current agricultural technologies to their best effect, getting out around 90 per cent of the theoretical maximum amount of food – and while this does involve a significant amount of soil erosion, there are alternative farming techniques that could minimise this without too much loss of yield. In America, meanwhile, that figure is only 70 per cent; in South America, Russia and Eastern Europe, it is under half; and in sub-Saharan Africa, less than a quarter.[95] As Fred Pearce writes in his book *Peoplequake*, which debunks many myths about overpopulation, 'Africa's problem is bad agriculture, not too many people'.[96]

It is true that water supplies may be a limiting factor on agricultural expansion, but Europe withdraws only 6 per cent of its renewable freshwater resource, and Asia only 20 per cent.[97] More to the point, only 11 per cent of global irrigation is delivered via sprinklers, which are far more efficient than traditional techniques – and only 1 per cent via drip irrigation, which is more efficient still.[98] The result, says McMahon, is that the current amount of 'crop per drop' is still disgustingly low, in large part because farmers do not have to pay the full cost of their water.

There is, in short, huge scope for expanding and refining current agricultural practices in order to feed and water the world – even with 80 million new mouths every year. Of course, there is the problem that the resources are often not in the right places, especially the water. That will require expensive infrastructure to fix. But these problems are not insuperable.

Even the 'food miles' decried by some environmentalists, who would prefer us all to eat only food grown as locally as possible, are something of a red herring. According to a study by the University of Wales Institute, shipping products from farm to grocery makes up just 2 per cent of their total environmental impact.[99] Furthermore, the modern market is so efficient that it treads lightly on the planet: it actually takes a firm like Wal-Mart less energy to move a giant truck filled with produce 300 miles than it does for pick-ups to take local produce to a farmers' market just 20 miles away.[100] Similarly, grass-fed New Zealand lamb and milk are massively less environmentally harmful than their grain-fed British equivalents, even after being shipped across the planet.[101]

ACCELERATION TO THE RESCUE

The situation becomes brighter still once you consider the possibilities for improving on our current techniques – which is a built-in imperative of an accelerated system, as it drives towards producing the maximum amount of food for the lowest energy input.

In Holland, they have put forward the idea of 'pig cities' – 76-storey farms in which the creatures can live in comfort before slaughter, with their manure powering a bio-gas digester. It has been calculated that you could feed the whole of London with 1,000 30-storey vertical farms, designed on similar principles.[102]

More important, as the information revolution reaches its culmination, food is becoming subject to the efficiencies of the supply chain in new ways. 'Produce firms on the cutting edge,' writes the entrepreneur Jeff Stibel, 'have sensors that measure various patches of soil for appropriate moisture and nutrients. Most send that information to the farmer; some send it straight to a fertilising robot that automatically spreads more water or fertiliser to that portion of soil.'[103]

Then there are the raw ingredients themselves. After long regulatory delays, 'golden rice' is finally making its way on to the world's dinner tables – enriched with beta-carotene, which the body converts into vitamin A, it has the potential to prevent millions of deaths a year from malnutrition and save hundreds of thousands of children in the developing world from preventable blindness.[104] In Israel, scientists have bred featherless chickens, which stay cooler in tropical climates and take up less room.[105] Other innovators are experimenting with offshore algae farms, which could suck carbon dioxide out of the atmosphere while providing energy-rich fertiliser, food and biofuels.[106]

The examples are legion. In the Netherlands, which is a hotbed of food-science research, Philips has developed special LEDs which give off precisely the right wavelengths of light craved by plants, making indoor farming far more effective.[107] Others have worked out how to irrigate crops with diluted seawater, after breeding a salt-tolerant potato.[108] It is now being planted across thousands of hectares in Pakistan, where 250 million people live on soil tainted with salt. In the US, meanwhile, scientists have developed 'heat-beater' beans that can thrive in temperatures up to 3°C higher than today's.[109]

All this comes before we have considered the possibility of more radical innovation – of applying the same disruptive spirit to food as we have to technology. We are already familiar, of course, with the idea of GM crops – which have now become a staple of the global shopping trolley. But their potential is still huge. The agri-giant Monsanto, a particular hate figure of the Green movement, believes that it can increase the number of bushels of corn per acre from 200 to 300 via GM, if not more.[110] A company called FuturaGene has already created a variety of eucalyptus that grows 40 per cent faster, for use in the timber trade and as a biofuel.[111] Scientists in Holland – again – have eliminated tomatoes' need for sleep, creating a variety that grows 24 hours a day, rather than needing 8 hours of darkness.[112] And genetic science of another sort has – thanks to the acceleration in computing power – made it far easier and cheaper to clone animals, ensuring that the healthiest and most productive specimens can be preserved and further enhanced.

But this is only the start. Across the world, tech companies and their founders are training their sights on the natural world. It was telling that when the first burger made from artificial beef, grown from cow stem cells at Maastricht University, was tucked into at a press conference in London in August 2013, it was revealed that the man behind the project was Sergey Brin, co-founder of Google.[113] Similarly, in China, Jack Ma, the charismatic founder of the internet giant Alibaba, has invested $320 million in an Inner Mongolian dairy company.[114] He is far from the only Chinese 'agri-tech' entrepreneur: the parent company of Lenovo, the world's biggest supplier of PCs, lists agriculture as one of its five key investment areas.[115] Such firms are dragging Chinese agriculture into the 21st century, and seeing yields skyrocket as a result.

Wherever you look, in fact, the new 'ag-tech' movement is applying the spirit of disruption to how we produce our food. Toshiba has converted an old disk-drive factory into a greenhouse producing 3 million heads of lettuce per year; it is kept so clean (the 'farmers' wear biohazard suits) that there is no need to wash dirt off the plants, or even use pesticides.[116] Caleb Harper at MIT has developed the 'CityFARM' project, in which sensors monitor each plant's demand for water, nutrients and carbon needs, as well as delivering the right wavelengths of light to maximise photosynthesis and flavour. This has cut water use by 98 per cent, quadrupled growth speed, eliminated fertilisers and pesticides, doubled the food's nutritional value and enhanced its flavour.[117]

This kind of farming can be done everywhere from skyscrapers to bomb shelters to shipping containers. As Harper told *Wired* magazine: 'Instead of someone picking a green tomato in Holland in the summer, or Spain in winter, with immature nutrients and a lack of flavour, and shipping it hundreds of kilometres, then gassing it with carbon dioxide so it turns red, you order your tomato and it's picked a few streets away and delivered to you fresh.'[118]

In Harper's vision, farming will divide between field-grown commodity crops and high-value perishables – perhaps grown in the back garden of the restaurant. Robot arms will scurry along, picking the fruit at just the right moment. In London, a company called Zero Carbon Food has converted old Tube tunnels into plant nurseries, using waste water in an 'ebb and flood' system, and growing the produce in pulverised carpet recycled from the Olympic village.[119]

Or perhaps, rather than growing better fruit and veg, we will dispense with them altogether. The *New Yorker* recently profiled the creator of Soylent – a food substitute that blends together all the essential vitamins and minerals and amino acids the body needs.[120] The product – named, in a macabre stroke, after the wonder-food in a Charlton Heston movie which turns out to be made from human beings – was created by Rob Rhinehart, who open-sourced his recipe and got crowd-funding to develop it further. It was not without its problems – when he put too little sodium into the blend, he felt 'foggy', and after overdoing it on the magnesium, he 'felt sharp pains throughout my entire body'. And it may well be that this craze peters out. But it does mesh neatly with the modern trend to treat meals as a brief necessity rather than a leisurely experience – and saves money that can be spent on going out to a restaurant for something more lavish.

Equally imaginative, but rather higher-powered, are firms such as Hampton Creek, which, rather than replacing traditional foods, want to reinvent food entirely. This San Francisco start-up is cataloguing all of the proteins found in the world's plants, analysing their make-up and behaviour and then recombining them to provide perfect facsimiles of existing products, or to create new foods entirely.[121] Backed by funding from Bill Gates and Li Ka-Shing, the Hong Kong billionaire, and led by YouTube's former chief data scientist, the firm's aim is to use big data to crunch through the entire plant kingdom and come up with alternatives to meat-based products and their messy and wasteful production. It has already devised a replica of a chicken egg, which is now used in a wide variety of products.

Such companies not only tend to be based in Silicon Valley, or backed by its leading figures, but to share the outsize, disruptive ambitions of the tech firms. Impossible Foods, one of Hampton Creek's rivals, promises to make 'beef that's better than any beef you've ever tasted' using proteins centrifuged from liquefied soy, wheat and spinach, juiced up with a special blend of amino acids and other chemical precursors of the aromas we associate with meat.[122] The aim is essentially to eliminate the cattle industry altogether – and help save the world by doing so.

'Our definition of success,' claims Pat Brown, the firm's founder, 'is that the world will look different from outer space, not that we see our label in Whole Foods.'[123] He is not alone in his optimism. Bill O'Neill, a pioneer of 3D-printing technology (who expects us to be able to print out many of these strange new nutritional concoctions in our own homes), predicts that the biophysicists behind projects like this will be 'the superheroes of this century'.[124]

We will not all be eating burgers made of plants any time soon. But such innovations – and dozens more – are evidence that the world is not standing still. The enormous acceleration of computing power has helped us understand biology and genetics with ever more accuracy, opening up scope for huge advances. For example, the reason beans or other legumes are always planted under crop-rotation systems is that they have a unique ability to fix nitrogen directly from the atmosphere back into the soil. When we fertilise crops we are artificially mimicking this process, using nitrogen created via the energy-intensive and carbon-emitting Haber-Bosch process. But if we could isolate the appropriate genes within the rhizobium bacteria – the symbiotic inhabitants of the root nodules of legumes – and transplant that ability into other crops, we would eliminate the need for artificial nitrogen at a stroke. That would save energy, avoid the pollution created by fertiliser run-off and help lower carbon emissions into the bargain.

THE FUTURE'S BRIGHT

There is no magic bullet that can solve all the world's ecological problems. Over the coming years, many people will continue to go hungry, and many aspects of the ecosystem will continue to be degraded. The world will also remain hugely unequal in how it shares out its natural blessings – or perhaps become even more so. But the very acceleration that got us into this mess is also the only way in which we can get out of it – not least by enabling our ability to feed ourselves to continue to outpace both our population and our increasingly sophisticated appetites.

Some of this will happen automatically, as a result of changed incentives. As scarcity develops, the higher prices will spur the development of alternatives, as has happened with energy. Competition will also see better and more efficient practices adopted. And as with food, so with the rest of the planetary system. In the words of the industrial analyst Peter Marsh: 'It is natural enough for people to worry that the products of manufacturing are contributing to the environmental problems that the world is struggling to counter. A truer depiction of the situation is not, however, that the world has too much manufacturing. It is that the world has too much of the wrong sort.'[125]

As the 'internet of things' develops, and the global supply chain grows more complex, more accurate and more self-aware, the scope for greater energy efficiency and reduced waste is huge. It is also worth noting that the more sophisticated the economy, the more efficient and less resource-intensive it becomes: for example, Chinese food production rose by almost 200 per cent between 1981 and 2007, while using only 50 per cent more fertiliser.[126] We may start to produce goods in our own homes, via 3D printing, eliminating huge amounts of waste – or to use more complicated and efficient nano-materials that can be recycled for other purposes.

In terms of energy, a suite of solar panels in the Sahara could provide for most of Europe's needs – alongside Icelandic volcanoes or Norwegian rivers.[127] If we can crack nuclear fusion, then we would end up with virtually unlimited, cheap, non-polluting energy – which could be used, among other things, to desalinate sea water and provide every house and every farm with all the water it required. There are also huge opportunities to capture energy used by trains or homes and return it to the grid.

Even humanity's greatest challenge of all – climate change – can be solved, or at least ameliorated, by our own ingenuity. There has been much debate recently in scientific and political circles about the ethics of 'geoengineering' – engaging in deliberate, as opposed to accidental, attempts to modify the contents of the planet's atmosphere. Ideas range from fertilising the oceans with iron filings to promote the growth of algae that mop up surplus carbon, to injecting soot into the stratosphere in order to cause the sun's rays to bounce back, to more outrageous schemes such as a network of giant orbital mirrors.

For some, this is unpardonable audacity – an inevitably ham-fisted attempt to atone for our own crimes with more ecological vandalism, which is bound to have unintended consequences. Others point out that we have already been making far more dramatic interventions by accident, so the case for making some by judicious design is extremely strong. It is notable that, in the range of forecasts the UN has produced to illustrate climate change's effects, the one that keeps warming below 2°C – the generally agreed safety threshold – already builds in the use of geoengineering on a massive scale, even though the technologies are currently unproven.[128]

It cannot be said strongly enough that geoengineering is very much a last-resort option. For one thing, it might encourage people to carry on polluting, in the belief that technology could sort the problem out. For another, the costs and benefits would – as ever – be unequally distributed.

Of all the plans proposed, injecting aerosols such as soot into the atmosphere is probably the simplest and most convincing. Yet even this would cool the Earth more at temperate (i.e. rich) latitudes and less at tropical (i.e. poor) ones.[129] It would alter the pattern of global weather, change the colour of the sky – and, if it were stopped, would cause the planet's temperatures to rebound rapidly to their original trajectory, leading to far more shocking transitions than under a more gradual progression. But we may not have a choice: taking such a gamble may be the only solution to our planetary short-termism that does not lead us to live on a sweltering planet.

If we are lucky, however, technological acceleration may provide alternative solutions. In New Jersey a researcher called Klaus Lackner has developed an artificial tree, whose densely packed leaves, coated in sodium carbonate resin, are 1,000 times more efficient than natural leaves at soaking up CO_2, which they store as baking soda.[130] By his calculations, 100 million of these trees – which would each cost the same as a car to mass-produce – could compensate for the entirety of humanity's carbon emissions. Make more, and more emissions would be absorbed. The by-products would be cooled and stored, then turned into eco-friendly fuel. At the moment, admittedly, the cost per tonne of CO_2 removed is uneconomical, but if humanity genuinely got serious about the problem, it is one among many potential solutions.

Ultimately, global warming – like most of the other side-effects of the great acceleration – may turn out to be what might be termed a 'fake bee problem'. I use this as a reference to the Harvard Microrobotics Lab, which as a solution to the decline in hive numbers, caused by the mysterious Colony Collapse Disorder, has produced tiny robots that can fulfil bees' economically vital pollinating function.[131]

For the Green lobby, it is a travesty that humanity has needed to come up with such a thing, purely because it has by its own ignorance endangered the system that nature created. And there's something in that. But the harsh truth is that, if we've inflicted the damage, it's better to have a fix than not. We may not find many of the solutions we come up with to our food needs, or our energy requirements, or our carbon problem, aesthetically appealing: who really wants to look at featherless chickens? But solutions will be found, and they will work – increasingly well, as the great acceleration delivers more advanced devices and technologies with which to devise them. We may only resort to them when all other options are exhausted, and when the immediate cost/benefit calculus finally tips in the right direction, but resort to them we will.

This is not to be Panglossian about the problems that we, as a species, face. As we have seen, many aspects of the great acceleration work against our long-term interests. In finance, there is a risk that short-termist investors lose patience before projects designed to solve the very biggest problems we face reach fruition. So there is obviously space for far-sighted and long-term government investment in projects such as nuclear fusion – which, depending on your point of view, is a decades-long, multi-billion-dollar money pit or quite the most exciting and transformative venture mankind has ever embarked upon.

Yet ironically, one of the benefits of the modern economy's tendency towards gigantism is that it has created Silicon Valley firms that are so large, and their owners so immune from shareholder pressure, that they can seriously consider investing equivalent resources to governments in 'moonshot' projects such as self-driving cars, space elevators or super-fast global Wi-Fi delivered via satellite or balloon – projects which not only benefit humanity, but push the great acceleration further on.

Better regulation can also help. Many of the problems we face have come about because capitalism has no inherent capacity to consider externalities – the environmental as well as physical cost of a product. There is, of course, a limited feedback loop when it comes to raw materials: as gold or neodymium or water become scarcer, their price will rise, changing producers' and consumers' behaviour. But there is no mechanism for a fertiliser company to tack on to the sale price the cost of cleaning up the nitrate run-off its products cause, or an oil major to pay for warming the planet by a trillionth of a degree with every tank of petrol it sells, or for a strawberry farmer in Spain or olive farmer in California to replenish the aquifers they are draining.

Tilting the tax system by making producers take more account of the full spectrum of costs of their goods, including the environmental ones, would not be anti-capitalist – it would make the system work properly, by ensuring that it was not slanted towards short-term exploitation of shared resources. It would also provide a more sustainable framework within which the economic imperatives of the great acceleration, and the disruptive influences of Silicon Valley, could do their work. Carbon taxes and emissions trading, for example, are market-based tools that are proven to cut emissions at low cost.[132]

Of course, such measures are electorally unpopular, because they ask people to pay now for the damage they will cause later – but as we have already seen, voters will come to terms with the pain surprisingly quickly. They are also vulnerable, in this global tragedy of the commons, to being undercut by 'free-riding' states that fail to impose such charges and profit from the results. Solving this problem is one of the great challenges facing our politics. Vertical farms or artificial meat will be delivered by the market as soon as they are cost-effective, even if it takes a bit of a shove to tax incentives to get there. But geoengineering, or solar farms in the Sahara, will require our fractious, short-termist politicians actually to sit down and talk to each other, something they have not had the greatest success at doing in the past.

Like so many of the other problems we face, our agricultural and environmental challenges are very far from insoluble. Our planet, and our species, are certainly in for a bumpy ride over the coming decades. But as Mark Lynas says in *The God Species*, his survey of our planetary impact, there is no convincing ecological reason 'why everyone in the world should not be able to enjoy rich-country levels of prosperity over the half-century to come'.[133] Yes, the relentless machinery of acceleration has caused many of our ecological problems – but it also offers us the tools to outgrow and out-think them. We just have to use those tools wisely.

Conclusion – Fast Forward

> 'Machines are to practically everybody what the white men were to the Indians.'
>
> – Kurt Vonnegut[1]

Throughout this book, we have explored the ways in which the great acceleration is transforming our society. In every sphere, it is disrupting our lives in ways both good and bad, bringing us new opportunities and fresh dangers all at the same time.

It is easy to be pessimistic about this process, to conclude that the bad outweighs the good – and certainly, its malign effects are more headline-grabbing. It is easier to picture a nightmarish stock market crash as a result of high-frequency trading than the quiet saving of fractions of a penny on billions of transactions, increasing the value of all of our savings and pensions, or to imagine our children's brains being destroyed by flashing lights than to accept that they are basically OK. But ultimately, acceleration is something we both desire and deserve – a shortcut to a richer, more convenient and more satisfying life for the great bulk of those rushed along in its wake.

Before we end, however, it is worth considering two final questions. First, is there anything that can disrupt the great acceleration, or bring it crashing to a halt? And second, where is it taking us? What does an accelerated future actually look like?

GREY PLANET

One of the most common responses to the great acceleration has been to call for a return to slowness – a yearning to go back to a time when life was simpler and more comprehensible, or at least to stop and catch our breath. Some, as we have seen, have suggested that an environmental collapse could force us to re-evaluate our priorities. Others warn of a long-term economic crisis, resulting from the way in which turbocharged capitalism chases its own tail, prioritising immediate profit over long-term prosperity. It is also possible that our own short-sighted short-termism will act as a brake on innovation: US investment in infrastructure, for example, is now at a 22-year low.[2]

Yet such an eventuality is deeply unlikely. As this book has shown, acceleration is so baked into the system that it will take an almighty alteration not just to our surroundings, but to our very biology, to deal it a serious blow. Even if the West were to retreat from haste and hustle, it is too late – the virus has escaped from its laboratory and infected the rest of the world with a desire to consume, innovate and disrupt.

Others, however, put forward another contender for acceleration's bane: old age. In sub-Saharan Africa the median age is currently barely 20. In India it is roughly 27. Yet in China it is a full decade higher, and in greying Japan and Germany it is a decade more than that.[3] Across the world, these figures are shooting upwards. An ageing society (which will also be a more female society, given that women tend to live longer than men) obviously has huge implications in terms of the cost of health and pension systems – implications with which the West is already struggling to come to terms.

But it will also affect what kind of people we are. An older planet might be less concerned with disruptive innovation and more concerned with security and familiarity – what Fred Pearce describes in his book *Peoplequake* as 'a stable, sagacious society that has lost its adolescent restlessness and settled into middle age'.[4] There is, indeed, intriguing evidence to back up this idea. The US biologist Robert Sapolsky carried out a series of experiments to determine how people's tastes change with age. He found that by our mid-thirties, and certainly by our forties, we lose our taste for new music, new cuisine and new habits more generally – we have found what we like, and tend to stick to it (something to bear in mind if your parents react with suspicion to culinary innovations such as sushi).[5]

Set against this, of course, are various countervailing trends. For one thing, it is possible that generations reared on disruption and acceleration will be more willing to update their tastes – doesn't every family have a story about the granny who took to her iPad, or Facebook, with alarming enthusiasm? Also, innovation is as often a function of collaboration as of individual insight – and there will be more brains than ever before who are capable of such collaboration, and trained to practise it. India, for example, is adding thousands if not millions of qualified engineers and entrepreneurs to the global network every year – and we have already seen that China is churning out PhDs as rapidly as it is iPhones.

With brainpower replacing muscle power as the key to economic success, such workers will have decades more to give to their work. And as Silicon Valley has amply demonstrated, it only takes a few driven and well-funded visionaries to upend industry after industry, and drive acceleration forward. What may change, however, will be the type of innovation: rather than messaging apps for teenagers, tomorrow's venture capitalists could be investing in personal care robots, telemedicine systems or exoskeletons to help the elderly walk.

A WORLD OF POSSIBILITIES

Even if human nature does somehow settle down, it will do nothing to slow the other force powering the great acceleration – the ever faster pace of technology. It is this that truly promises (or threatens) to reshape our lives, to an extent that we can as yet only barely appreciate. When we think of the future, we often imagine it as a speedier version of the present-day West. A more likely scenario is that it will look like nothing we can imagine.

The process of acceleration has thus far been concentrated in the field of information technology and productivity – which has already played out across the rest of the economy and society in a hugely powerful fashion. But what happens when that same amount of raw computing power and data-processing knowhow is applied to biology, or materials science?

Just as we can sift through data, breaking it down into its component parts, so information technology could soon grant us the ability to break down our genetic code, or even matter itself. As with data, we could then rebuild it into new and exotic arrangements. Superconductive graphene, transparent aluminium, synthetic elements, structures that can repair and replicate themselves, nanomaterials – all promise to transform the world around us, as well as our impact upon it.

As for biology, it is now possible – thanks to the acceleration of computing power – to carry out the same level of genetic sequencing in a garage (or by sending samples away in the post) that once required a laboratory packed with the world's most powerful computers.[6] J. Craig Venter, creator of Synthia, the world's first artificial bacterium, has been explicit about his desire to turn biology into engineering – to remix living organisms into better and more convenient shapes.[7] We already have created goats that excrete spider silk in their milk. 'BioBricks' – Lego-like biological building blocks of DNA that can be snapped together to produce the chosen enzymes or proteins – offer near-infinite opportunities for further customisation. One configuration already proposed, if sent in a vial to Mars, would be able to turn its soil into compacted construction material.[8]

As with the other facets of the great acceleration, it is already possible to see how these developments might spur on others, many of them unpleasant and alarming. Instructions for using 3D printing to produce your own handgun from a bag of plastic or metal powder have been downloaded thousands of times, despite the best efforts of law enforcement.[9] The price of a printer capable of building the gun is at around £10,000 at the time of writing, but it will be far cheaper even by the time this book is on the shelves. And the USA has already embarked on a covert programme to protect the president's DNA from being collected by foreign powers, and to gather genetic material from his counterparts overseas.[10] It may not be long before an occupant of the White House (or a disgruntled employee at a biotech firm) can deliver a viral payload precision-engineered to target Vladimir Putin, or a particularly irksome boss.

Such genetic hacking may seem far-fetched, but consider the following alarming thought experiment. One of the more enlightened new autocracies – Singapore seems a good example – announces that its scientists have isolated a set of genes they believe are responsible for human intelligence. From now on, embryos will be screened to eliminate those in which the right genes are not switched on – or, as genetic science advances, tweaked to ensure that they are. Imagine the headlines about a nation of Einsteins – and the pressures for other countries, or just individual parents, to follow suit for fear of being left behind in the genetic arms race.

You do not even need to advance genetic science much beyond its current state: a new and extraordinarily powerful technology called CRISPR allows you to edit the human or animal genome as easily as a Word document. Bioengineers have already used it to copy and paste genes to allow yeast to consume plant matter and excrete ethanol – a genuine and potentially planet-saving biofuel. But they have also used it in China to alter non-viable human embryos, in an attempt (as yet unsuccessful) to correct a particular genetic mutation.[11]

Genetic engineering does not even need to go that far to have an effect. Even without sculpting new and better human beings from scratch, it could be theoretically possible to select the 'best' combinations of sperm and egg out of, say, 20 samples from two given parents, then cultivate further sperms and eggs from these and match them to others. Within a few genetic generations, you could effectively have a version of humanity with all the flaws ironed out.

This prospect is as frightening as it is exciting – not least because of how it would turn the existing divide between haves and have-nots into a yawning chasm. Yet the consolation would be to imagine what such superhumans could do when armed with all the engineering, digital and genetic technology which would have appeared in the interim. 'Consider,' says Nick Bostrom (of whom more below), 'how the rate of progress in the field of artificial intelligence' – for example – 'would change in a world where Average Joe is an intellectual peer of Alan Turing or John von Neumann.'[12]

At the moment, of course, it is the digital revolution that is the focus of most of our concerns – in particular the idea that computers can do our jobs better, faster and more cheaply. In technical terms, this is known as 'capital-biased technological change' – the fact that the traditional division of income between capital and labour is inexorably tilting in favour of the former, in the shape of the people who own the best computers and fastest machines.[13] Honda recently opened an ultra-sophisticated, ultra-automated factory in its home territory of Japan. If its techniques were rolled out across the entire global car industry, the number of employees who would be needed to produce the same number of cars would be under a million, an order of magnitude lower than the current figure.[14]

As the sophistication of robots and algorithms grows exponentially, more and more industries face being swallowed up – shelf-stackers, composers, surgeons, accountants, journalists. Many now suggest that such innovations, while enormously beneficial for humanity as a whole, will destroy more jobs than they create – Foxconn, the Chinese company that makes the iPad and thousands of other goods, recently announced a 100-fold expansion in its use of robotics, since its human workers were getting dangerously expensive.[15] 'Baxter', a new industrial robot in the US, can carry out the same tasks as humans – its creators claim – for as little as $3 an hour.[16] Tech start-ups are wonderful things, but they employ precious few people when compared with old-school factories.

Even in the most optimistic version of this, in which automation opens up industries and opportunities that we cannot imagine, there is still likely to be a shift away from big mass employers and towards a more piecemeal, scrappy and uncertain life of freelance creative work or even domestic servitude (on the assumption that there will be a lingering cachet, among the new ruling class, in having old-fashioned human servitors or creators at their disposal). And as we have seen, politics and government will inevitably lag behind in their responses to such changes.

As a result, the divide between those empowered by this new lifestyle, and those left behind, is likely to grow still further. This, in turn, means more political dislocation, especially given the potential of these new technologies to make the world more fragile and frenetic even as they make it more wondrous.

There are already those who argue that London, or Silicon Valley, or Google should pull an Ayn Rand and secede from the unworthy world that surrounds them: the nation state, they argue, is running on a buggy, outdated codebase and needs to be rebooted. That may not be likely in the immediate, or even the medium term. But it will take imaginative and far-sighted policymakers to come up with ways to share the fruits of disruption equally enough that techno-Utopia for the few does not become dystopia for the many.

THE LAST INVENTION

Lurking behind this acceleration in computing capability is a more revolutionary prospect still – that the Anthropocene may give way, in surprisingly short order, to the Technocene, as we fuse with, or are replaced by, artificial minds and bodies. This will be via the development of what some call 'the last invention' – an artificial brain clever enough to upgrade itself, at which point we humans become rather irrelevant to the process.

For Ray Kurzweil, the path to this point (which he calls the Singularity) will probably be via the development of the 'nonbiological neocortex' – a device either embedded in our bodies, or more likely as part of the cloud, that augments our brainpower with computing power. 'We have already outsourced much of our personal, social, historical and cultural memory to the cloud,' he argues, 'and we will ultimately do the same with our hierarchical thinking.'[17]

To Kurzweil, the future is a marvellous place, in which intelligent nanobots swim through our brains and bloodstream, eliminating flaws and upgrading performance. Already, he points out, we have created devices the size of blood cells that can detect and destroy cancer cells, or produce insulin to counteract the effects of diabetes. Sonia Trigueros, a nanobiologist at the Oxford Martin School, has built new nanoscale antibacterials, as well carbon nanotubes that deliver chemotherapy drugs directly into the cells – killing 95 per cent of cancerous cells in just 25 hours.[18]

If exponential improvements in computer power are reflected in this field too, we can expect such medical devices to be a billion times more powerful within 30 years – along with the rest of our devices.[19] At that point, there may be little holding us back from the ultimate upgrade – to switch from biological bodies into digital minds, ideally made of 'computronium', a hypothetical substance in which the atoms are arranged so as to foster calculation at the fastest speeds materially possible. 'Over time,' gushes Kurzweil, 'we will convert much of the mass and energy in our tiny corner of the galaxy that is suitable for this purpose to computronium' – then all we need to worry about are the limitations of the speed of light as we spread ourselves to the stars.[20]

Others, however, are less sanguine about the challenges ahead. Professor Martin Rees, Britain's former Astronomer-Royal, is so concerned that he and other big thinkers (including one of the co-founders of Skype) have set up a new institute at Cambridge University: the Centre for the Study of Existential Risk.[21] They worry that biotechnology, nanotechnology and artificial intelligence create the potential for things to go so wrong, so fast and on such a scale that humanity will be unable to react in time: the equivalent of a Flash Crash in the markets, but in the real world.

This may sound far-fetched, but the scenarios they conjure are eerily realistic. There is the possibility of a super-plague, engineered either by a terrorist group, a disaffected lab worker or even a naive idealist (such as the scientists at the University of Wisconsin-Madison who artificially recreated a virus almost identical to the Spanish flu that killed tens of millions of people in the wake of the First World War, then attempted to publish their results).[22] There is the possibility of catastrophic climate change, or of one or more nations experimenting with geoengineering and having it go horribly wrong. And there is what is known as the 'paperclip scenario'.

This thought experiment is put forward by Nick Bostrom, who has been in the forefront of those warning that artificial intelligence would bring catastrophic dangers as well as vast opportunities.[23] It holds, essentially, that it is highly unlikely that the first artificial intelligence will be anything like those we are familiar with from science fiction – whether a helpful, human-sounding C3PO type like Tony Stark's sidekick JARVIS, or an evil, megalomaniacal force like *Terminator*'s Skynet. Instead, its intelligence is unlikely to resemble anything human at all.

Why does that matter? Because such entities will look at the world in a very different way from us. Let us say you set such an AI a simple task: its core programming will be to maximise the number of paperclips produced, or to calculate the value of pi, or solve the Riemann hypothesis. In pursuit of this, the AI's logical course of action would be to maximise the resources available to it – even though they may already be in use by, or even consist of, human beings.

If we are in its way, it could produce a self-replicating nano-nerve gas which burgeons forth from every square metre of the globe. Or, says Bostrom:

> if the AI is sure of its invincibility to human interference, our species may not be targeted directly. Our demise may instead result from the habitat destruction that ensues when the AI begins massive global construction projects using nanotech factories and assemblers – construction projects which quickly, perhaps within days or weeks, tile all of the Earth's surface with solar panels, nuclear reactors, supercomputing facilities with protruding cooling towers, space rocket launchers, or other installations whereby the AI intends to maximise the cumulative long-term realisation of its values.[24]

Once the entire solar system has been turned into computronium – or paperclips – the AI could then seed the stars with von Neumann probes containing its own code, and replicate itself endlessly across the cosmos. Even a seemingly benign command – for example, to maximise human happiness – could have unintended consequences, as the AI simply wired us all up to experience a constant sensation of neurological bliss: the ultimate extension of *The Matrix*, but at our own misguided request.

This may sound alarmist, or ludicrous, but Bostrom and others in the field make an extremely convincing case for why we should be alarmed. True AI – which we can define as a computer with the ability and the tools to upgrade itself – is likely to come about in an 'arms race' environment, or at the very least one in which the primary aim will be to get it to work first and worry about the consequences later.

The prospect of controlling the world's first superintelligence is attractive enough to prompt anyone to cut corners – and one can see why financial firms, or national militaries, would be especially concerned to develop artificial strategists that can out-think their opponents, and devise better versions of themselves faster than any rivals. But even if such entities appeared to be friendly when confined to 'black boxes' in which they cannot affect the outside world, once they were unleashed into the wild we would have no way of coping – given their enormously superior intelligence and capability – if or when the time came for them to pursue their own ends rather than ours.

A genuine AI is also likely to think and act with a speed that humans cannot match – just as rogue algorithms can play havoc in the financial markets before humans have had time to push the stop button. As Bostrom points out, while we think of idiots and geniuses as wildly different in terms of intelligence, they are actually pretty close together in the grand scheme of things: Einstein and your typical reality TV contestant are both brainy when compared with a dog or lizard, and dumb when compared with an AI that has developed the ability to get smarter at the speed of silicon rather than evolution.[25]

Even if – as Kurzweil predicts – such beings emerge from a fusion of human and machine intelligence, rather than a pure AI, they might rapidly evolve into something completely different and distinct. Consider a just-smarter-than-human mind, set loose in a digital wonderland of processing power. It would be a trivial matter to generate copies of itself – perhaps a dozen, perhaps a million – and devote them to devising ways to improve its own intelligence, or the intelligence of the resulting collective. Humanity could, says Bostrom, find itself 'deposed from its position of apex cogitator over the course of an hour or two'.[26]

For Bostrom, 'before the prospect of an intelligence explosion, we humans are like small children playing with a bomb. Such is the mismatch between the power of our plaything and the immaturity of our conduct . . . what we have here is not one child but many, each with access to an independent trigger mechanism. The chances that we will ALL find the sense to put down the dangerous stuff seem almost negligible.'[27] As a result, he views the development of superintelligence as 'quite possibly the most important and most daunting challenge humanity has ever faced' – and quite probably the last, since it will, for good or ill, take such decisions out of our hands.[28]

It may come to pass, of course, that there is some insuperable barrier that prevents computers achieving true sentience, or even the appearance of it – but then, as we have seen in the financial markets, intelligence is not a prerequisite of wreaking algorithmic havoc.

Certainly, today's best efforts are laughably primitive. Honda's multi-million-dollar Asimo robot, for example, can mimic certain human movements – but only by wandering around in a creepy half-crouch. It can also bring you a drink – but only if you spend quite a while pre-programming it with the location of both the bar and your table.[29]

Yet even if computers remain our tools, they will be tools of enormous power. IBM's primitive AI Watson, which was designed to win the quiz show *Jeopardy*, is now being used in medical diagnosis, with several hundred versions running on its servers, all working in parallel.[30] A study in *Wired* found that private investment in AI has been expanding by 62 per cent a year – and it will soon be used to power Amazon's recommendations, or Google's search engine. According to this report:

> Like all utilities, AI will be supremely boring, even as it transforms the Internet, the global economy, and civilization. It will enliven inert objects, much as electricity did more than a century ago. Everything that we formerly electrified we will now cognitize.

> This new utilitarian AI will also augment us individually as people (deepening our memory, speeding our recognition) and collectively as a species.[31]

Bostrom's warnings may not come to pass – AI may enter our lives gradually rather than explosively, as a helpful and welcomed friend who becomes more and more capable at smoothing the path of our lives without ever overstepping his bounds. But we are duty-bound, if there is the slimmest possibility of any of his scenarios occurring, to devote the necessary amount of time and mental ingenuity to staving them off. For whether AI is friendly or hostile, its impact will be so vast precisely because it will be accelerated. As more of us use these new computing services, they will grow smarter, leading more people to use them: a virtuous cycle that embodies precisely the same tendency to gigantism that we have seen elsewhere.

It is for this reason that the possibility of an 'intelligence explosion', of whatever kind, deserves so much attention. Whatever the means by which it comes about – biological, technological, or some mixture of the two – if an entity or group is able to increase its intelligence beyond the human baseline, even by a small amount, then it can in theory build on that advantage, speeding away from the rest of us more rapidly than we can catch up. That may happen over decades or over minutes. But either way, it would still have hugely disruptive consequences.

And yet, all we have seen so far shows that we will almost certainly not stop to think, or pause to consider, before plunging into the creation of such super-intelligences. Making computers faster and our bodies better delivers immediate rewards – maybe not to all of us, but certainly to vast numbers of consumers and a small number of entrepreneurs who own the devices and processes involved. And if there is one thing that has been consistent across the great acceleration, from finance to our social habits to the state of the environment, it has been our willingness to discount future risks in pursuit of present convenience. This would be the ultimate expression of that trend, in every sense.

Wherever you look in our society, things are changing – fast. The East is rising and the West is fading. Old certainties and habits are being discarded. The digital elite proclaim their intention to bring disruption to our homes, to our jobs, to our food, to our genes and ultimately to life and death themselves – and by and large, we cheer them on for it.

Think back to the image at the start of this book, of the feet marching ever faster through our great new cities. Are they striding confidently into a prosperous future, or scurrying to keep up? Are they getting faster because they want to, or because they have to? The answer, of course, is both. The nature of 21st-century society, its most basic setting, is to accelerate – which is why speed is, and will remain, the most important force in our lives. This process will inevitably bring terrors as well as wonders: the price of a faster pace is a bumpier ride. Yet on balance, an accelerated future is one we will surely rush to embrace.

Acknowledgements

First of all, thanks to you, the reader, for sticking with me this long. (Unless you're a friend of mine who's skipped to the back to see whether they're mentioned, in which case shame on you.) I hope this book was as interesting to read as it was to write.

Secondly, thanks to the many people who agreed to be interviewed, either on or off the record, and the many more whose books, articles and research I relied upon. I don't have the space to list more than a fraction of them, but if you visit thegreatacceleration.com you'll find reading lists for each of the topics covered here, as well as new articles. Please do come along and say hi.

For a book about speed, this took an agonisingly long time to write. It would not exist without the help of many people: Michael Fishwick and all at Bloomsbury, who took a chance on a novice author and then bore with him with great patience; Elly James, who was the first to believe in the book, before handing the baton on to her colleague Heather Holden-Brown; Andrew Wright, who brought order to the chaos of my initial draft; and Julia Kingsford, Henry Volans, Luke McGee, Harry Wilson, Angela Lascelles, Jon Coleman and the others who read various chapters and versions and provided encouragement and criticism when each was needed.

Many other people helped me out in ways large and small: Ed Howker, Mark Forsyth, Neil O'Brien, Max Pemberton, David Bodanis, the Colvile, Coleman and Pelham families, the rest of my long-suffering friends. Thank you to all my colleagues at the *Telegraph* and BuzzFeed for the opportunities they gave me and the inspiration they provided: Liz Hunt, Roger Highfield, Iain Martin, Fraser Nelson, Ben Brogan, Tony Gallagher, Bryony Gordon and so many, many more. In particular, this book wouldn't exist without Chris Deerin, Sally Chatterton and the rest of the team on the *Telegraph* comment desk, whose relentless mockery spurred me to finally write the damn thing.

There are so many other people I could and should thank, so if you haven't been mentioned yet but feel you ought to have been, please consider me appropriately grateful. But finally and above all, thanks to my son Edward (whose contribution to this project has been necessarily limited) and to his mother Andrea. Rare is the wife who'll proofread every page of her husband's book. Rarer still is the one who'll put up with him writing a chapter of it on their honeymoon. I'm very lucky, and very grateful.

Notes

INTRODUCTION

1 Umberto Boccioni *et al.*, 1910, http://www.unknown.nu/futurism/techpaint.html.
2 Carl Honoré, *In Praise of Slow*, Orion, London, 2005, p. 7.
3 Robert Levine, *A Geography of Time*, Basic Books, New York, 1997, p. 16.
4 Frank Partnoy, *Wait*, PublicAffairs, Philadelphia, 2012, p. 104.
5 Levine, *Geography*, p. 131.
6 See http://www.richardwiseman.com/quirkology/pace_home.htm.
7 See for example John Freeman, *The Tyranny of E-mail*, Scribner, New York, 2009, p. 183.
8 Stefan Klein, *The Secret Pulse of Time*, Da Capo Press, London, 2007, p. 164.
9 Ibid., p. 154.
10 Danny Dorling, *So You Think You Know About Britain?*, Constable & Robinson, London, 2011, p. 46.
11 Stephanie Brown, *Speed*, Berkley, New York, 2014, p. 6.
12 Ibid., p. 64.
13 Brigid Schulte, *Overwhelmed*, Bloomsbury, London, 2014, p. 44.
14 Jonathan Franzen, 'Rage Against the Machine', *Guardian*, 13 Sept 2013 (no longer available online).
15 Ryan Kelly, 'How Webpage Load Time Is Related to Visitor Loss', Pear Analytics, Aug 2009, http://pearanalytics.com/blog/2009/how-webpage-load-time-related-to-visitor-loss/.
16 See https://developers.google.com/speed/docs/insights/Server.
17 Philip Zimbardo & John Boyd, *The Time Paradox*, Rider, London, 2010, p. 324.
18 John Maynard Keynes, 'Economic Possibilities for our Grandchildren', 1930, http://www.econ.yale.edu/smith/econ116a/keynes1.pdf.
19 Levine, *Geography*, p. 10.

20 The Economist, *Megachange*, Profile, London, 2012, p. 202.
21 Ibid., p. 155.
22 Adam Gopnik, 'The Information', *New Yorker*, Feb 2011, http://www.newyorker.com/magazine/2011/02/14/the-information.
23 XKCD, 'The Pace of Modern Life', https://xkcd.com/1227/.
24 Freeman, *Tyranny*, p. 56.
25 Ibid., p. 49.
26 Ibid., p. 47.
27 James Gleick, *Faster*, Random House, New York, 1999.
28 Karl Marx, *Manifesto of the Communist Party*, chapter 1, https://www.marxists.org/archive/marx/works/1848/communist-manifesto/.
29 Freeman, *Tyranny*, p. 198.

1 PERMANENT REVOLUTION

1 Meeting attended by the author.
2 See e.g. Chris Nuttall, 'Valley-backed Color's "miraculous" New App', *Financial Times* Tech Blog, 24 March 2011, http://blogs.ft.com/tech-blog/2011/03/valley-backed-colors-miraculous-new-app/.
3 Martin Hilbert & Priscila Lopez, 'The World's Technological Capacity to Store, Communicate, and Compute Information', *Science*, 2011, http://www.uvm.edu/~pdodds/files/papers/others/2011/hilbert2011a.pdf.
4 Larry Downes, *The Laws of Disruption*, Basic Books, New York, 2009, p. 11.
5 See http://www.kurzweilai.net/the-law-of-accelerating-returns.
6 James Gleick, *The Information*, Fourth Estate, London, 2011, p. 395.
7 All statistics from the author's visit to CERN.
8 Usman Pirzada, 'Intel ISSCC: 14nm All Figured Out, 10nm Is On Track, Moore's Law Still Alive and Kicking', *WCCFtech*, 5 Feb 2015, http://wccftech.com/intel-isscc-14nm/.
9 Stephen Levy, 'Google's Larry Page on why Moonshots Matter', *Wired*, 17 Jan 2013, http://www.wired.com/2013/01/ff-qa-larry-page/.
10 John Freeman, *The Tyranny of E-mail*, Scribner, New York, 2009 p. 36.
11 Ibid., p. 38.
12 For further details see http://www.orphanspreferred.com.
13 Freeman, *Tyranny*, p. 39.
14 Tom Standage, *The Victorian Internet*, Weidenfeld & Nicolson, London, 1998.
15 Ibid., p. 58.
16 Ibid.
17 Ibid., p. 79.

18 Ibid., p. 77. Rather embarrassingly, the cable was so poorly made that it promptly broke; it took almost a decade for a permanent link to be established.
19 Freeman, *Tyranny*, p. 46.
20 Stephen Kern, *The Culture of Time and Space, 1880–1918*, Harvard University Press, Harvard, 2003, p. 70.
21 Robert Penn, *It's All About the Bike*, Penguin Travel, London, 2011, p. 3.
22 Kern, *Culture*, p. 111.
23 Joseph A. Schumpeter, 'Creative Destruction', 1942 https://notendur.hi.is/~lobbi/ut1/a_a/SCUMPETER.pdf.
24 For the full details see http://www.pewresearch.org/fact-tank/2014/03/14/chart-of-the-week-the-ever-accelerating-rate-of-technology-adoption/.
25 Andrew Keen, *Digital Vertigo*, Constable & Robinson, London, 2012, p. 30.
26 Brian Winston, *Media, Technology and Society*, Routledge, New York, 1998, p. 7.
27 Niccolò Machiavelli, *The Prince*, chapter VI, https://www.gutenberg.org/files/1232/1232-h/1232-h.htm.
28 Winston, *Media*, pp. 89 & 259.
29 Tim Wu, *The Master Switch*, Knopf, New York, 2010.
30 Clayton M. Christensen, *The Innovator's Dilemma*, Harvard Business Review Press, Harvard, 1997.
31 The Economist, *Megachange*, Profile, London, 2012, p. 195.
32 American Enterprise Institute, 'Charts of the Day: Creative Destruction in the S&P 500 Index', 26 Jan 2014, https://www.aei.org/publication/charts-of-the-day-creative-destruction-in-the-sp500-index/.
33 Toni Mack, 'Danger: Stealth Attack', *Forbes*, 25 Jan 1999, http://www.forbes.com/forbes/1999/0125/6302088a.html.
34 Jocelyn R. Davis, Henry M. Frechette Jr and Edwin H. Boswell, *Strategic Speed*, Harvard Business Press, Harvard, 2010, p. 9.
35 John Cassidy, *Dot.con*, HarperCollins, New York, 2003, p. 254.
36 Google executive in meeting attended by the author.
37 See Gina Keating, *Netflixed*, Penguin, New York, 2013.
38 See https://en.wikipedia.org/wiki/Netflix_Prize.
39 John Palfrey & Urs Gasser, *Born Digital*, Basic Books, New York, 2008, p. 230.
40 David Kirkpatrick, *The Facebook Effect*, Virgin Books, London, 2010, p. 59.
41 Jeremy Rifkin, *The Zero Marginal Cost Society*, Palgrave Macmillan, New York, 2014, p. 129.
42 Karl Marx, *Manifesto of the Communist Party*, chapter 1, https://www.marxists.org/archive/marx/works/1848/communist-manifesto/.

43 Marc Andreeson, 'Why Software Is Eating The World', *Wall Street Journal*, 20 Aug 2011, http://www.wsj.com/articles/SB10001424053111903480904576512250915629460.

44 Tad Friend, 'Tomorrow's Advance Man', *New Yorker*, 18 April 2015, http://www.newyorker.com/magazine/2015/05/18/tomorrows-advance-man.

45 Ibid.

46 Ludwig Siegele, 'A Cambrian Moment', *Economist*, 18 Jan 2014, http://www.economist.com/news/special-report/21593580-cheap-and-ubiquitous-building-blocks-digital-products-and-services-have-caused.

47 In 2014 Mark Zuckerberg claimed that in Facebook's maturity this had been adapted to 'Move fast with stable infra', which isn't quite as sexy. Samantha Murphy Kelly, 'Facebook Changes Its "Move Fast and Break Things" Motto', *Mashable*, 30 April 2014, http://mashable.com/2014/04/30/facebooks-new-mantra-move-fast-with-stability/.

48 Matt Warman, 'Can The Web Make The World Go Faster?', 18 Nov 2010, *Telegraph*, http://www.telegraph.co.uk/technology/facebook/8140562/Can-the-web-make-the-world-go-faster.html.

49 BBC News, '"Fail Fast" Advises LinkedIn Founder and Tech Investor Reid Hoffman', 11 Jan 2011, http://www.bbc.co.uk/news/business-12151752.

50 Douglas Coupland, 'Why McLuhan's Chilling Vision Still Matters Today', *Guardian*, 20 July 2011, http://www.theguardian.com/commentisfree/2011/jul/20/marshall-mcluhan-chilling-vision.

51 Liz Gannes, 'Lyft and Uber Price Wars Leave Some Drivers Feeling Crunched', *Re/Code*, 30 April 2014, http://recode.net/2014/04/30/lyft-and-uber-price-wars-leave-some-drivers-feeling-crunched/.

52 Joseph A. Schumpeter, *Capitalism, Socialism and Democracy*, Taylor & Francis e-Library, 2003, p. 32, http://digamo.free.fr/capisoc.pdf.

53 Philip Zimbardo & John Boyd, *The Time Paradox*, Rider, London, 2010, pp. 40–1.

54 Laura M. Holson, 'Putting a Bolder Face on Google', *New York Times*, 1 March 2009, http://www.nytimes.com/2009/03/01/business/01marissa.html.

55 Brian Christian, 'The A/B Test: Inside the Technology That's Changing the Rules of Business', *Wired*, 25 April 2012, http://www.wired.com/2012/04/ff_abtesting/.

56 Technically Google is now a subsidiary of a holding company called Alphabet, but I have used the same name for both to avoid confusion, and because everyone else does.

57 See https://www.youtube.com/watch?v=aXJklICrFJI.

58 See https://www.google.co.uk/about/company/philosophy/.

59 Jake Brutlag, 'Speed Matters', Google Research blog, 23 June 2009, http://googleresearch.blogspot.co.uk/2009/06/speed-matters.html.

60 See https://www.youtube.com/watch?v=aXJklICrFJI.

61 Kohavi, Longbotham, Sommerfeld & Henne, 'Controlled Experiments on the Web: Survey and Practical Guide', *Data Mining And Knowledge Discovery*, Springer, 2008, http://ai.stanford.edu/~ronnyk/2009controlledExperimentsOnTheWebSurvey.pdf.

62 Meeting attended by author.

63 'Search: Now Faster Than the Speed of Type', Google official blog, 8 Sept 2010, http://googleblog.blogspot.co.uk/2010/09/search-now-faster-than-speed-of-type.html.

64 Kirkpatrick, *Facebook*, pp. 70–6.

65 Gary Rivlin, 'Wallflower at the Web Party', *New York Times*, 15 Oct 2006, http://www.nytimes.com/2006/10/15/business/yourmoney/15friend.html.

66 Kirkpatrick, *Facebook*, p. 58.

67 Ibid., p. 82.

68 Ibid., p. 61.

69 Ibid., p. 192.

70 Ibid., p. 303.

71 Emma Barnett, 'Facebook Launches Next Generation of Email', *Telegraph*, 15 Nov 2010, http://www.telegraph.co.uk/technology/facebook/8135442/Facebooks-Mark-Zuckerberg-launches-next-generation-of-email.html.

72 IFPI Digital Music Report 2011, http://www.ifpi.org/content/library/DMR2011.pdf.

73 Study no longer online but summarised here: http://readwrite.com/2010/10/06/study_pirated_E-books_on_the_rise.

74 Business Sofware Alliance, 'Software Piracy on the Internet: A Threat To Your Security', Oct 2009, http://portal.bsa.org/internetreport2009/2009internetpiracyreport.pdf.

75 Downes, *Disruption*, p. 252.

76 Intellectual Property Office, 'Online Copyright Infringement Tracker, Wave 5', 22 June 2015, https://www.gov.uk/government/uploads/system/uploads/attachment_data/file/449592/new_OCI_doc_290715.pdf.

77 Ryan Faughnder, 'Music Piracy Is Down But Still Very Much in Play', *Los Angeles Times*, 20 June 2015, http://www.latimes.com/business/la-et-ct-state-of-stealing-music-20150620-story.html.

78 Farhad Manjoo, 'Will Netflix Destroy The Internet?', *Slate*, 20 Nov 2010, http://www.slate.com/articles/technology/technology/2010/11/will_netflix_destroy_the_internet.html.
79 Brad Stone, *The Everything Store*, Transworld, London, 2014, p. 100.
80 Ibid., p. 365.
81 Ibid., p. 302.
82 '40% of Amazon Items Now Come from Third Party Sellers', Digital Strategy Consulting, http://www.digitalstrategyconsulting.com/intelligence/2015/01/40_of_amazon_items_now_come_from_third_party_sellers.php.
83 Stone, *Everything Store*, p. 232.
84 Ron Amadeo, 'Cheaper Bandwidth or Bust: How Google Saved YouTube', ArsTechnica, 23 April 2015, http://arstechnica.com/gadgets/2015/04/cheaper-bandwidth-or-bust-how-google-saved-youtube/.
85 Kirkpatrick, *Facebook*, p. 210.
86 Stone, *Everything Store*, p. 23.
87 Miguel Helft, 'Google's Larry Page: The Most Ambitious CEO in the Universe', *Fortune*, 13 Nov 2014, http://fortune.com/2014/11/13/googles-larry-page-the-most-ambitious-ceo-in-the-universe/.
88 Nick Winfield & Brian Stelter, 'How Netflix Lost 800,000 Customers And Goodwill', *New York Times*, 25 Nov 2011, http://www.nytimes.com/2011/10/25/technology/netflix-lost-800000-members-with-price-rise-and-split-plan.html.
89 Farhad Manjoo, 'Is Netflix As Dumb As It Seems?', 19 Sept 2011, *Slate* http://www.slate.com/articles/technology/technology/2011/09/is_netflix_as_dumb_as_it_seems.html.
90 Elaine Moore, 'Cross-Border Capital Flows Return to 2011 Levels', *Financial Times*, 30 Nov 2014, http://www.ft.com/cms/s/0/10803656–74b5–11-e4–8321–00144feabdc0.html.
91 Statistics from *OECD Yearbook 2012* via http://www.oecdobserver.org/news/fullstory.php/aid/3681/An_Emerging_middle_class.html.

2 QUICK REACTIONS

1 Jonathan Franzen, 'Rage Against the Machine', *Guardian*, 13 Sept 2013 (no longer available online).
2 George Miller Beard, *American Nervousness: Its Causes and Its Consequences*, G. P. Putnam, New York, 1881. Reprinted on demand by General-Books.net but page numbers unavailable.
3 Ibid.
4 Ibid.

5 Ibid.
6 John Freeman, *The Tyranny of E-mail*, Scribner, New York, 2009, p. 78, & Stephen Kern, *The Culture of Time and Space, 1880–1918*, Harvard University Press, Harvard, 2003, p. 125.
7 Brigid Schulte, *Overwhelmed*, Bloomsbury, London, 2014, p. 48.
8 Susan Greenfield, *2121*, Head of Zeus, London, 2013, p. 83.
9 Ibid., p. 3.
10 Ibid., p. 4.
11 Philip Zimbardo & John Boyd, *The Time Paradox*, Rider, London, 2010, p. 322.
12 Walter E. Houghton, *The Victorian Frame of Mind 1830–1870*, Yale University Press, Yale, 1957, p. 7.
13 Nicholas Carr, *The Shallows*, Atlantic Books, 2010, p. 5.
14 David J. Linden, *Pleasure*, Oneworld, Oxford, 2011, p. 3.
15 Ibid., p. 9.
16 Ibid., p. 14.
17 Christian Rudder, *Dataclysm*, Fourth Estate, London, 2014, p. 216.
18 Carr, *Shallows*, p. 116.
19 'Everything is Connected', *Economist*, 5 Jan 2013, http://www.economist.com/news/briefing/21569041-can-internet-activism-turn-real-political-movement-everything-connected.
20 Micheline Maynard & Matthew L Wald, 'Off-Course Pilots Cite Computer Distraction', *New York Times*, 27 Oct 2009, http://www.nytimes.com/2009/10/27/us/27plane.html.
21 Carr, *Shallows*, p. 121.
22 Geoffrey Miller, 'Why We Haven't Met Any Aliens', *Seed*, 30 April 2006, http://seedmagazine.com/content/article/why_we_havent_met_any_aliens/.
23 Daniel Levitin, *The Organized Mind*, Dutton, New York, 2014, p. 170.
24 Kathryn Mills, 'Effects of Internet Use on the Adolescent Brain', *Trends in Cognitive Sciences*, 18(8), 2014.
25 Jonah Lehrer, *Imagine*, Canongate, Edinburgh, 2012, p. 62.
26 Clive Thompson, *Smarter Than You Think*, William Collins, London, 2013, p. 133.
27 Stephen Levy, 'This Is Your Brain on Twitter', *Backchannel*, 5 Feb 2015, https://medium.com/backchannel/this-is-your-brain-on-twitter-cac0725cea2b.
28 Carr, *Shallows*, p. 118.
29 Freeman, *Tyranny*, p. 12.
30 Research by IDC for Facebook, summarised at http://www.slideshare.net/jeffrufino/idc-study-mobile-and-social-connectiveness.

31 See for example Freeman, *Tyranny*, p. 134.
32 Michael Y. Park, 'Cell Phones Are The Latest "Addiction"', *Fox News*, 18 July 2006, http://www.foxnews.com/story/2006/07/18/cell-phones-are-latest-addiction.html.
33 Freeman, *Tyranny*, p. 138.
34 Levitin, *Organized*, p. 170.
35 Schulte, *Overwhelmed*, p. 63.
36 Ibid., p. 63.
37 David Pierce, 'iPhone Killer: The Secret History of the Apple Watch', *Wired*, April 2015, http://www.wired.com/2015/04/the-apple-watch.
38 Thompson, *Smarter*, p. 136.
39 Schulte, *Overwhelmed*, p. 68.
40 Daniel Goleman, *Focus*, Bloomsbury, London, 2013, p. 9.
41 Ibid., p. 7.
42 Tony Dokoupil, 'Is The Internet Making Us Crazy?', *Newsweek*, 7 Sept 2012, http://www.newsweek.com/internet-making-us-crazy-what-new-research-says-65593.
43 Ibid.
44 Maggie Jackson, *Distracted*, Prometheus Books, New York, 2009, p. 18.
45 Michael S. Rosenwald, 'Serious Reading Takes a Hit from Online Scanning and Skimming, Researchers Say', *Washington Post*, 6 April 2014, http://www.washingtonpost.com/local/serious-reading-takes-a-hit-from-online-scanning-and-skimming-researchers-say/2014/04/06/088028d2-b5d2–11e3-b899–20667de76985_story.html.
46 Matthew Hutson, 'People Prefer Electric Shocks to Being Left Alone With Their Thoughts', *The Atlantic*, 3 July 2014, http://www.theatlantic.com/health/archive/2014/07/people-prefer-electric-shocks-to-being-alone-with-their-thoughts/373936/.
47 Tom Cheshire, 'Distracted? How Hyperstimulation Is Making You Smarter', *Wired*, 10 Dec 2013, http://www.wired.co.uk/magazine/archive/2013/12/features/hyperstimulation/viewgallery/330469.
48 Susan Cain, *Quiet*, Penguin, London, 2012, p. 85.
49 Levitin, *Organized*, p. 170.
50 Damian Thompson, *The Fix*, Collins, London, 2012, p. 3.
51 Goleman, *Focus*, p. 56.
52 Ibid., p. 31.
53 Daniel Kahneman, *Thinking, Fast and Slow*, Penguin, London, 2011, p. 41.
54 Ibid., p. 44.
55 Sendhil Mullainathan & Eldar Shafir, *Scarcity*, Allen Lane, London, 2013, p. 30.
56 Goleman, *Focus*, p. 220.

57 Ibid., p. 43.
58 Zimbardo & Boyd, *Time Paradox*, p. 324.
59 Claudia Hammond, *Time Warped*, Canongate, Edinburgh, 2012, p. 62.
60 Zimbardo & Boyd, *Time Paradox*, p. 16.
61 For example, between 1950 and the late 1980s the proportion of US teens who said they considered themselves 'an important person' rose from 12 per cent to 80 per cent. David Brooks, *The Social Animal*, Short Books, London, 2011, p. 191.
62 See for example Howard Gardner and Katie Davis, *The App Generation*, Yale University Press, Yale, 2013.
63 Emily Dugan, 'Going to Work Is More Stressful than Ever, Poll Reveals', *Independent*, 2 Nov 2014, http://www.independent.co.uk/life-style/health-and-families/health-news/going-to-work-is-more-stressful-than-ever-poll-reveals-9833602.html.
64 Freeman, *Tyranny*, p. 161.
65 Schulte, *Overwhelmed*, p. 22.
66 Robert M. Sapolsky, *Monkeyluv*, Vintage, 2006, p. 86. This autonomic nervous system also recovers more quickly in men than women, meaning that a wife may still be agitated about an argument or accident well after her husband has forgotten about it.
67 Ibid., p. 17.
68 Ibid., p. 94.
69 Schulte, *Overwhelmed*, p. 59.
70 Nilesh J. Samani & Pim van der Harst, 'Biological Ageing and Cardiovascular Disease', *Heart*, 2008.
71 Danny Dorling, *So You Think You Know About Britain?*, Constable & Robinson, London, 2011, p. 46.
72 Schulte, *Overwhelmed*, p. 60.
73 Bill Hathaway, 'Even In The Healthy, Stress Causes Brain To Shrink, Yale Study Shows', *Yale News*, 1 Sept 2012, http://news.yale.edu/2012/01/09/even-healthy-stress-causes-brain-shrink-yale-study-shows.
74 Schulte, *Overwhelmed*, p. 62.
75 Po Bronson & Ashley Merryman, *Top Dog*, Ebury Press, London, 2013, p. 68.
76 Ibid., p. 69.
77 Ibid., p. 75.
78 Ibid., p. 77.
79 Ibid., p. 103.
80 Ethan Watters, *Crazy Like Us*, Constable & Robinson, London, 2011, p. 213.
81 Schulte, *Overwhelmed*, p. 61.

82 Ibid., p. 30.
83 Ibid., p. 27.
84 Ibid., p. 50.
85 Jodi Kantor & David Streitfeld, 'Inside Amazon: Wrestling Big Ideas in a Bruising Marketplace', *New York Times*, 16 Aug 2015, http://www.nytimes.com/2015/08/16/technology/inside-amazon-wrestling-big-ideas-in-a-bruising-workplace.html.
86 Emma Jacobs, 'Workaholic Ex-Bankers Impose Long-Hours Culture on their Colleagues', *Financial Times*, 21 March 2013, http://www.ft.com/cms/s/0/3c26b148-ae9c-11e3-aaa6–00144feab7de.html.
87 Stefan Klein, *The Secret Pulse of Time*, Da Capo Press, London, 2007, p. 217.
88 Interview with the author.
89 Schulte, *Overwhelmed*, p. 22.
90 Sarah Boseley, 'Working Longer Hours Increases Stroke Risk, Major Study Finds', *Guardian*, 20 Aug 2015, http://www.theguardian.com/lifeandstyle/2015/aug/20/working-longer-hours-increases-stroke-risk.
91 Keith Perry, 'Stress Can Be Transmitted Through TV Screen', *Telegraph*, 1 May 2014, http://www.telegraph.co.uk/news/science/10802814/Stress-can-be-transmitted-through-TV-screen.html.
92 Schulte, *Overwhelmed*, p. 64.
93 Ibid., p. 139.
94 Levitin, *Organized*, p. 307.
95 Schulte, *Overwhelmed*, p. 88.
96 Ibid., p. 33.
97 Ibid.
98 Mullainathan & Shafir, *Scarcity*, p. 155.
99 Schulte, *Overwhelmed*, p. 207.
100 Richard Layard, *Happiness*, Penguin, London, 2011, p. 150.
101 Mullainathan & Shafir, *Scarcity*, p. 58.
102 Maria Konnikova, 'Why Can't We Fall Asleep?', *New Yorker*, 7 July 2015, www.newyorker.com/science/maria-konnikova/why-cant-we-fall-asleep.
103 Rachel Carlyle, 'The Real Reason Were All So Tired', *Times Weekend*, 28 Sept 2013, http://www.thetimes.co.uk/tto/health/article3880963.ece.
104 Konnikova, 'Why Can't We Fall Asleep?'
105 Russell Foster, interview with the author.
106 Steve Brown, interview with the author.
107 For example in Till Roenneberg *et al.*, 'Social Jetlag and Obesity', *Current Biology*, May 2012.
108 Interview with the author.
109 International Agency for Research on Cancer press release, 5 Dec 2007, http://www.iarc.fr/en/media-centre/pr/2007/pr180.html.

110 Clare Anderson *et al.*, 'Deterioration of Neurobehavioral Performance in Resident Physicians During Repeated Exposure to Extended Duration Work Shifts', *Sleep*, Aug 2012, http://www.ncbi.nlm.nih.gov/pubmed/22851809. The brain-scanning experiment, using a smaller number of subjects, was described by Charles Czeisler of Harvard in a public lecture.
111 Carl Honoré, *In Praise of Slow*, Orion, London, 2005, p. 7.
112 Maria Konnikova, 'The Walking Dead', *New Yorker*, 9 July 2015, http://www.newyorker.com/science/maria-konnikova/the-walking-dead.
113 Roenneberg *et al.*, 'Light and the Human Circadian Clock', *Handbook of Experimental Pharmacology*, Jan 2013.
114 Interview with the author.
115 Konnikova, 'The Walking Dead'.
116 Interview with the author.
117 Konnikova, 'The Walking Dead'.
118 Eve Van Cauter *et al.*, 'The Impact of Sleep Deprivation on Hormones and Metabolism', *Medscape Neurology*, 2005.
119 Interview with the author.
120 Christina M. Kelly, 'Lights Out, Phones On', *NBC News*, 1 Nov 2010, http://www.nbcnews.com/id/39917869/ns/health-childrens_health/t/lights-out-phones-many-teens-text-all-night-long/.
121 Interview with the author.
122 'Blue Light Has a Dark Side', *Harvard Health Letter*, 1 May 2012, http://www.health.harvard.edu/staying-healthy/blue-light-has-a-dark-side.
123 Gregory Ferenstein, 'How I Keep Screens from Causing Me Insomnia', *The Verge*, 24 March 2015, http://www.theverge.com/2015/3/24/8278221/screen-use-insomnia-sleep-disruption-sunglasses.
124 Frances Booth, *The Distraction Trap*, Pearson, Harlow, 2013, p. 142.
125 Hannah Richardson, 'Later School Start Time "May Boost GCSE Results"'. *BBC News*, 9 Oct 2014, http://www.bbc.co.uk/news/education-29461685.
126 See for example Richard Thaler & Cass Sunstein, *Nudge*, Penguin, London, 2009.
127 Thompson, *Smarter*, p. 220.
128 Tanjil Rashid, 'Swedish City Embarks on Six-Hour Workday Experiment', *Telegraph*, 9 April 2014, http://www.telegraph.co.uk/news/worldnews/europe/sweden/10754656/Swedish-city-embarks-on-6-hour-workday-experiment.html.
129 Schulte, *Overwhelmed*, p. 136.
130 'Work Less', Treehouse official blog, 7 March 2011, http://blog.teamtreehouse.com/work-less.
131 Schulte, *Overwhelmed*, p. 139.

132 Michael Arrington, 'Startups Are Hard, So Work More, Cry Less and Quit All the Whining', *Uncrunched*, 27 Nov 2011, http://uncrunched.com/2011/11/27/startups-are-hard-so-work-more-cry-less-and-quit-all-the-whining/.
133 Schulte, *Overwhelmed*, p. 266.
134 Sapolsky, *Why Zebras Don't Get Ulcers: An Updated Guide To Stress, Stress-Related Diseases, and Coping*, Barnes & Noble Books, New York, 2000.
135 Goleman, *Focus*, p. 47.
136 Ibid., p. 81.
137 Ibid., p. 78.
138 Ibid., p. 83.
139 Jackson, *Distracted*, p. 23.
140 Ibid.
141 Ibid.
142 Frank Partnoy, *Wait*, PublicAffairs, Philadelphia, 2012, p. 4.
143 Goleman, *Focus*, p. 187.
144 Ibid., p. 193.
145 Partnoy, *Wait*, p. 14.
146 Goleman, *Focus*, p. 196.
147 Schulte, *Overwhelmed*, p. 208.
148 Ibid.
149 Ibid., p. 60.
150 Britta K. Hölzel *et al.*, 'Mindfulness Practice Leads to Increases in Regional Brain Gray Matter Density', *Psychiatry Research*, Jan 2011.
151 Hölzel *et al.*, 'Stress Reduction Correlates with Structural Changes in the Amygdala', *Social Cognitive and Affective Neuroscience*, Sept 2009.
152 Levitin, *Organized*, p. 306.
153 Martin Rossman, *The Worry Solution*, Rider Books, London, 2010, p. 172

3 FAST FRIENDS

1 Lauren Collins, 'The Love App', *New Yorker*, 25 Nov 2013, http://www.newyorker.com/magazine/2013/11/25/the-love-app.
2 'Oh! You Pretty Things', *Economist*, 10 July 2014, http://www.economist.com/news/briefing/21606795-todays-young-people-are-held-be-alienated-unhappy-violent-failures-they-are-proving.
3 Ibid.
4 Interview with the author. He is technically now executive chairman of Google's parent company Alphabet, but I have kept his old title for simplicity's sake.
5 David Brooks, *The Social Animal*, Short Books, London, 2011.

6 Brigid Schulte, *Overwhelmed*, Bloomsbury, London, 2014, p. 26.

7 John Bingham, 'Forget Affairs – How Office Email is the New "Third Person" in Britain's Marriages', *Telegraph*, 16 Feb 2014, http://www.telegraph.co.uk/women/sex/divorce/10638975/Forget-affairs-how-the-office-email-is-the-new-third-person-in-Britains-marriages.html.

8 In a study cited by Maggie Jackson, only 17 per cent of families consistently ate dinner with each other. Maggie Jackson, *Distracted*, Prometheus Books, New York, 2009, p. 63.

9 In the same study, stay-at-home wives stopped what they were doing to welcome their husbands back from work a little more than a third of the time. Most children didn't even look up. Jackson, *Distracted*, p. 62.

10 Aric Sigman, 'Time For A View On Screen Time', *Archives of Disease in Childhood*, Oct 2012.

11 Frances Booth, *The Distraction Trap*, Pearson, Harlow, 2013, p. 10.

12 Andrew Keen, *Digital Vertigo*, Constable & Robinson, London, 2012, p. 67.

13 Schulte, *Overwhelmed*, p. 240.

14 Ibid.

15 Richard Watson, *Future Minds*, Nicholas Brealey, London, 2010, p. 29.

16 Jackson, *Distracted*, p. 73.

17 Ibid.

18 Watson, *Future Minds*, p. 29.

19 David L. Gilden and Laura R. Marusich, 'Contraction of Time in Attention-Deficit Hyperactivity Disorder', *Neuropsychology*, 2009.

20 Fifty-four per cent of children aged between four and six preferred to watch TV than spend time with their father. Watson, *Future Minds*, p. 29.

21 Public Health England, 'How Healthy Behaviour Supports Children's Wellbeing', 28 Aug 2013, https://www.gov.uk/government/uploads/system/uploads/attachment_data/file/232978/.

22 Booth, *Distraction*, p.9.

23 Daniel Levitin, *The Organized Mind*, Dutton, New York, 2014, p. 368.

24 David Pakman, 'May I Have Your Attention, Please?', *Medium*, 10 Aug 2015, https://medium.com/life-learning/may-i-have-your-attention-please-19ef6395b2c3.

25 Alex Kantrowitz, 'Parents Are Punishing Their Kids By Making Them Watch TV', BuzzFeed, 8 July 2015, http://www.buzzfeed.com/alexkantrowitz/parents-are-punishing-their-kids-by-making-them-watch-tv.

26 Aleks Krotoski, *Untangling the Web*, Faber & Faber/Guardian Books, London, 2013, p. 174.

27 Stephen Levy, 'This Is Your Brain on Twitter', *Backchannel*, 5 Feb 2015, https://medium.com/backchannel/this-is-your-brain-on-twitter-cac0725cea2b.

28 Malcolm Gladwell claims the average person under 25 texts as much in a day as an over-55 in a year. Other statistics show a tenfold difference. John Koetsier, 'Malcolm Gladwell: the Snapchat Problem, the Facebook Problem, the Airbnb Problem', 24 July 2015, http://venturebeat.com/2015/07/24/gladwell-on-data-marketing-the-snapchat-problem-the-facebook-problem-the-airbnb-problem/.

29 Howard Gardner and Katie Davis suggest generations should now be defined by the technologies they use, and therefore turn over much more rapidly. Gardner & Davis, *The App Generation*, Yale University Press, Yale, 2013, pp. 13–14.

30 Sherry Turkle, *Alone Together*, Basic Books, New York, 2011.

31 Ibid., p. xii.

32 Neil Howe, 'Why Millennials Are Texting More and Talking Less', *Forbes*, 15 July 2015, http://www.forbes.com/sites/neilhowe/2015/07/15/why-millennials-are-texting-more-and-talking-less/.

33 Turkle, p. 2.

34 Krotoski, *Untangling*, p. 83.

35 Dan Slater, *Love in the Time of Algorithms*, Current, New York, 2013, p. 178.

36 Ibid., p. 183.

37 According to the latest Pew research, 42 per cent of social media-using teens have had someone post things about them that they cannot change or control; 21 per cent have felt worse about their own life because of what they see from other friends; 40 per cent feel pressure to post only content that makes them look good; and 88 per cent believe people share too much information about themselves. Amanda Lenhart, 'Teens, Technology and Friendships'. Pew Research Center, 6 Aug 2015, http://www.pewinternet.org/2015/08/06/teens-technology-and-friendships/.

38 Krotoski, *Untangling*, p. 103.

39 US college students are now in contact with their parents 13.4 times a week on average. Gardner & Davis, *App Generation*, p. 85.

40 Krotoski, *Untangling*, p. 89.

41 Interview with the author.

42 Rosalind Wiseman, 'What Boys Want', *Time*, 2 Dec 2013, http://time.com/696/what-boys-want/.

43 Philip Larkin, 'This Be the Verse', *High Windows*, Faber & Faber, London, 1974.

44 Schulte, *Overwhelmed*, p. 207.

45 danah boyd, *It's Complicated*, Yale University Press, London, 2014, p. 88.

46 Roger Mackett *et al*, 'Children's Independent Movement in the Local Environment', *Built Environment*, 1990, http://www.ucl.ac.uk/~ucetjop/pdf/benv.33.4.454.pdf.

47 Gardner & Davis, *App Generation*, p. 94.
48 boyd, *Complicated*, p. 88.
49 Ibid., p. 121.
50 Jackson, *Distracted*, p. 22.
51 Henry Hitchings, *The Language Wars*, John Murray, London, 2011, p. 293.
52 Ibid., p. 295.
53 See Christian Rudder, *Dataclysm*, Fourth Estate, London, 2014, pp. 59–62.
54 Clive Thompson, *Smarter Than You Think*, William Collins, London, 2013, p. 66.
55 Rebecca Solnit, 'Diary', *London Review of Books*, 29 Aug 2013, http://www.lrb.co.uk/v35/n16/rebecca-solnit/diary.
56 Charlotte Philby, 'Pro-Suicide Websites Pushed My Teenage Son to Take His Own Life', *Independent*, 14 Aug 2014, http://www.independent.co.uk/life-style/gadgets-and-tech/features/prosuicide-websites-pushed-my-teenage-son-to-take-his-own-life-9669944.html.
57 Rudder, *Dataclysm*, p. 235.
58 Jamie Bartlett, *The Dark Net*, William Heinemann, London, 2014, p. 170.
59 See https://en.wikipedia.org/wiki/2014_celebrity_photo_hack.
60 Bartlett, *Dark Net*, pp. 15–19.
61 Ben Hammersley, *64 Things You Need to Know Now for Then*, Hodder & Stoughton, London, 2012, p. 47.
62 See http://www.penny-arcade.com/S=0/comic/2004/03/19.
63 Christopher Steiner, *Automate This*, Portfolio/Penguin, New York, 2012, p. 145.
64 Levitin, *Organized*, p. 130.
65 'E-Daters Picky About Partners', *Times Higher Education*, 15 April 2005, https://www.timeshighereducation.co.uk/news/e-daters-picky-about-partners/195362.article.
66 James Vincent, 'WhatsApp Evidence Used to Divorce Nearly Half of Italian Adulterers', *Independent*, 10 Nov 2014, http://www.independent.co.uk/life-style/gadgets-and-tech/whatsapp-evidence-used-to-divorce-nearly-half-of-italian-adulterers-9850780.html.
67 Krotoski, *Untangling*, p. 104.
68 Rudder, *Dataclysm*, p. 123.
69 Ibid., p. 103.
70 Ibid., p. 89.
71 Slater, *Algorithms*, p. 75.
72 Ibid., p. 76.
73 Steiner, *Automate*, p. 145.
74 Amy Webb, *Data, A Love Story*, Dutton, New York, 2013, pp. 188–200.

75 Kevin Poulsen, 'How a Maths Genius Hacked OKCupid to Find True Love', *Wired*, 21 Jan 2014, http://www.wired.com/2014/01/how-to-hack-okcupid/.
76 Dan Ariely, *The Upside of Irrationality*, HarperCollins, London, 2010, p. 220.
77 Ibid., p. 222.
78 Nancy Jo Sales, 'Tinder and the Dawn of the "Dating Apocalypse"', *Vanity Fair*, Sept 2015, http://www.vanityfair.com/culture/2015/08/tinder-hook-up-culture-end-of-dating.
79 Jessica Massa, *The Gaggle*, Simon & Schuster, New York, 2013.
80 Sales, 'Tinder'.
81 Sarah Knapton, 'Sex Will Soon Be Just for Fun Not Babies, Says Father of the Pill', *Telegraph*, 9 Nov 2014, http://www.telegraph.co.uk/news/health/news/11217750/Sex-will-soon-be-just-for-fun-not-babies-says-father-of-the-Pill.html.
82 V. S. Ramachandran, *The Tell-Tale Brain*, William Heinemann, London, 2011, p. 74.
83 Slater, *Algorithms*, p. 11.
84 Ibid., p. 121.
85 See https://en.wikipedia.org/wiki/Ashley_Madison_data_breach.
86 Slater, *Algorithms*, pp. 121–2.
87 Annalee Newitz, 'Almost None of the Women in the Ashley Madison Database Ever Used the Site', *Gizmodo*, 26 Aug 2015, http://gizmodo.com/almost-none-of-the-women-in-the-ashley-madison-database-1725558944.
88 John Bingham, 'The Family Is More Popular Than Ever', *Telegraph*, 26 March 2013, http://www.telegraph.co.uk/news/uknews/9953364/The-family-is-more-popular-than-ever.html.
89 Louisa Peacock & Laura Donnelly, 'Women Delaying Motherhood Is "Worrying Issue", Says Chief Medical Officer', *Telegraph*, 16 Jan 2014, http://www.telegraph.co.uk/women/womens-health/10578227/Women-delaying-motherhood-is-worrying-issue-says-Britains-chief-doctor.html.
90 Ed Howker & Shiv Malik, *Jilted Generation*, Icon, London, 2010, p. 63.
91 Slater, *Algorithms*, p. 119.
92 Sales, 'Tinder'.
93 Divorce rates in the UK, for example, have fallen sharply since their high in the early 1990s. See http://www.ons.gov.uk/ons/rel/vsob1/divorces-in-england-and-wales/2012/info-divorces.html.
94 Mark Penn, *Microtrends*, Allen Lane, London, 2007, p. 23.
95 Amanda Hess, 'Such Tweet Sorrow', *Slate*, 8 April 2014, http://www.slate.com/articles/double_x/doublex/2014/04/study_says_twitter_causes_infidelity_and_divorce_don_t_believe_it.html.

96 Harry Fisch, *The New Naked*, Sourcebooks, Illinois, 2013.
97 *Grazia* magazine, not available online.
98 Collins, 'The Love App'.
99 Tyler Cowen, *Average Is Over*, Dutton, New York, 2013, p. 73.
100 Natasha Lamas, 'Save a Swipe – This Bot Selects Tinder Dates for You', *TechCrunch*, 10 Feb 2015, http://techcrunch.com/2015/02/10/tinderbox/.
101 Jay Ferrari, 'How to Become a Human Sex Toy', *Men's Health*, 26 Feb 2015, http://www.menshealth.com/best-life/vibrating-penis-implants?page=7.
102 Max Fisher, 'Japan's Sexual Apathy Is Endangering the Global Economy', *Washington Post*, 22 Oct 2013, https://www.washingtonpost.com/news/worldviews/wp/2013/10/22/japans-sexual-apathy-is-endangering-the-global-economy/.
103 Roberto A. Ferdman, 'Americans Aren't Getting Married, and Researchers Think Porn Is Part of the Problem', *Washington Post*, 21 Dec 2014, http://www.washingtonpost.com/blogs/wonkblog/wp/2014/12/21/americans-arent-getting-married-and-researchers-think-porn-is-part-of-the-problem/.
104 Turkle, *Alone Together*, chapter 2.
105 David Levy, *Love + Sex With Robots*, HarperCollins, 2007, p. 249.
106 Ibid.
107 Interview with the author.
108 Robert Putnam, *Bowling Alone*, Simon & Schuster, New York, 2001, p. 331.
109 Henry Hemming, *Together*, John Murray, London, 2011, p. 146.
110 Thompson, *Smarter*, p. 283.

4 THE ART OF ACCELERATION.

1 Interview with *PBS Newshour*, 10 Nov 2004, transcript available at http://www.pbs.org/newshour/bb/entertainment-july-dec04-roth_11-10/.
2 See http://www.spritzinc.com.
3 David Gritten, '*Jurassic Park 3D* Review', *Telegraph*, 22 Aug 2013, http://www.telegraph.co.uk/culture/film/filmreviews/10259543/Jurassic-Park-3D-review.html.
4 A. O. Scott, 'The Death of Adulthood in American Culture', *New York Times*, 11 Sept 2014, http://www.nytimes.com/2014/09/14/magazine/the-death-of-adulthood-in-american-culture.html.
5 Interview with the author.
6 James E. Cutting *et al.*, 'Quicker, Faster, Darker: Changes in Hollywood Film Over 75 Years', *iPerception*, 2011.

7 Ibid.

8 Outlined in a series of posts at http://blog.echonest.com in late 2013. Search for the tag 'audio trends over time'.

9 Glenn McDonald, 'From Elvis to Miley, "Danceability" Remains Constant', Echo Nest, 3 Sept 2013, http://blog.echonest.com/post/60178016824/from-elvis-to-miley-danceability-remains.

10 Glenn McDonald, 'Is Music Getting More Energetic Over Time?', Echo Nest, 10 Sept 2013, http://blog.echonest.com/post/60842073276/is-music-getting-more-energetic-over-time.

11 Author's figures, using data from http://www.billboard.com/charts/hot-100.

12 Neil McCormick, 'Top 100 Music Downloads: "Pure, Escapist Entertainment"', *Telegraph*, 21 April 2014, http://www.telegraph.co.uk/culture/music/rockandpopmusic/10778071/Top-100-music-downloads-pure-escapist-entertainment.html.

13 John Seabrook, 'The Doctor Is In', *New Yorker*, 14 Oct 2013, http://www.newyorker.com/magazine/2013/10/14/the-doctor-is-in.

14 James Masterton, 'Sales of the Century', *Popbitch*, 25 Feb 2015, http://popbitch.com/home/2015/02/25/sales-of-the-century/.

15 Bob Lefsetz, 'Why The Grammys Matter', *Lefsetz.com*, 29 Jan 2015, http://lefsetz.com/wordpress/index.php/archives/2015/01/29/spotifymediabase-triple/.

16 Mark Mulligan, 'What Future for the Album in the On-Demand Age?', *Music Industry Blog*, 15 July 2014, https://musicindustryblog.wordpress.com/2014/07/15/what-future-for-the-album-in-the-on-demand-age/.

17 Interview with the author.

18 Bryan Appleyard, 'Forty Years of *The Sunday Times* Bestsellers', *Sunday Times*, 13 April 2014, http://www.thesundaytimes.co.uk/sto/culture/books/non_fiction/article1397632.ece.

19 Mark Penn, *Microtrends*, Allen Lane, London, 2007, p. 197.

20 Boris Kachka, 'When Did Books Get So Freaking Enormous? The Year of the Very Long Novel', *Slate*, 29 May 2015, http://www.slate.com/blogs/browbeat/2015/05/29/when_did_books_get_so_freaking_Enormous_the_year_of_the_very_long_novel.html.

21 See http://blog.smashwords.com/2014/07/2014-smashwords-survey-reveals-new.html, slide 32 in the embedded Powerpoint presentation.

22 Mark Coker, 'Do E-book Customers Prefer Longer or Shorter Books?', *Huffington Post*, 26 June 2012, http://www.huffingtonpost.com/mark-coker/do-ebook-customers-prefer_b_1457011.html.

23 Steven Johnson, *Everything Bad Is Good for You*, Penguin, London, 2006, p. 67.

24 Matt Zoller Seitz, 'The Rise of Fast TV: Why *Empire* Is the Preferred Script Model', Vulture, May 17 2015, http://www.vulture.com/2015/05/rise-of-fast-tv.html.
25 Clarissa Tan, 'Is Sherlock Starting To Suffer From ADHD?', *Spectator*, 4 Jan 2014, http://www.spectator.co.uk/arts/television/9104342/thank-heavens-sherlock-aka-benedict-cumberbatch-is-not-dead/.
26 Andrew Harrison, 'Steven Moffat: "I Was the Original Angry *Doctor Who* Fan"', *Guardian*, 18 Nov 2013, http://www.theguardian.com/tv-and-radio/2013/nov/18/steven-moffat-doctor-who-interview.
27 Daniel Levitin, *This Is Your Brain on Music*, Atlantic Books, 2008.
28 David Brooks, *The Social Animal*, Short Books, London, 2011, p. 349.
29 Todd VanDerWerff, 'Why *Louie* Is the Next Stage in the Evolution of the TV Sitcom', *AV Club*, 27 Sept 2012, http://www.avclub.com/article/why-ilouie-iis-the-next-stage-in-the-evolution-of--85474.
30 Charles Duhigg, *The Power of Habit*, William Heinemann, London, 2012, pp. 197–209.
31 Ibid., p. 199.
32 Ibid., p. 208.
33 Ibid.
34 Chris Anderson, *The Long Tail*, Random House, New York, 2006.
35 Anita Elberse, *Blockbusters*, Faber & Faber, London, 2014.
36 Ibid., p. 161.
37 Ibid.
38 Derek Thompson, 'The Shazam Effect', *Atlantic*, Dec 2014, http://www.theatlantic.com/magazine/archive/2014/12/the-shazam-effect/382237/.
39 Elberse, *Blockbusters*, p. 20.
40 Gerry Smith, 'TV's "Golden Age" Won't Last Because You're Not Watching Enough', *Bloomberg*, 28 May 2015, http://www.bloomberg.com/news/articles/2015-05-28/tv-s-golden-age-won-t-last-because-you-re-not-watching-enough.
41 Elberse, *Blockbusters*, p. 67.
42 Ibid., p. 68.
43 Ibid., p. 23.
44 Ibid., p. 65.
45 John B. Thompson, *Merchants of Culture*, Polity Press, Cambridge, 2012, p. 269.
46 Ibid.
47 Interview with the author.
48 Thompson, *Merchants*, p. 223.
49 Ibid., p. 224.

50 John McDuling, 'Hollywood Is Giving Up On Comedy', *Atlantic*, 3 July 2014, http://www.theatlantic.com/entertainment/archive/2014/07/the-completely-serious-decline-of-the-hollywood-comedy/373914/.

51 Elberse, *Blockbusters*, p. 20.

52 John Palfrey & Urs Gasser, *Born Digital*, Basic Books, New York, 2008, p. 230.

53 Joe Miller, 'Microsoft Pays $2.5 Billion for *Minecraft* Maker Mojang', 15 Sept 2014, *BBC News*, http://www.bbc.co.uk/news/technology-29204518.

54 Cecilia Kang, 'TV Is Increasingly for Old People', *Washington Post*, 5 Sept 2014, http://www.washingtonpost.com/news/business/wp/2014/09/05/tv-is-increasingly-for-old-people/.

55 Jonathan Ford, 'The Diary: Jonathan Ford'. *Financial Times*, 29 Aug 2014, http://www.ft.com/cms/s/0/fe598136–2e9a-11e4-bffa-00144feabdc0.html.

56 The video has been removed for violating YouTube's terms of service, but can still be found fairly easily.

57 Cecilia Kang, 'The Real Reasons Why YouTube's 5 Biggest Stars Became Millionaires', *Washington Post*, 23 July 2015, https://www.washingtonpost.com/news/the-switch/wp/2015/07/23/how-these-5-youtube-stars-became-millionaires-and-why-you-wont-be-joining-them-anytime-soon/.

58 Ibid.

59 Wesley Yin-Poole, 'PewDiePie Talks Money', Eurogamer, 8 July 2015, http://www.eurogamer.net/articles/2015–07–08-pewdiepie-talks-money-haters-gonna-hate.

60 Max Benator, 'Can You Still Become YouTube Famous?', *Re/Code*, 5 May 2015, http://recode.net/2015/05/05/can-you-still-become-youtube-famous/.

61 Pew Research Center, 'Teen Content Creators and Consumers', Nov 2005, http://www.pewinternet.org/2005/11/02/teen-content-creators-and-consumers/ & Jeff Howe, *Crowdsourcing*, Random House, London, 2009, p. 270.

62 Tim Lewis, 'YouTube Superstars: The Generation Taking on TV – and Winning', *Observer*, 7 April 2013, http://www.theguardian.com/technology/2013/apr/07/youtube-superstars-new-generation-bloggers.

63 Ford, 'The Diary'.

64 Ellen Cushing, 'Sexts, Hugs and Rock and Roll', *BuzzFeed*, 2 July 2015, http://www.buzzfeed.com/ellencushing/sexts-hugs-and-rock-n-roll-on-the-road-with-the-teen-social.

65 Lewis, 'YouTube Superstars'.

66 Thompson, *Merchants*, p. 382.

67 Taylor Swift, 'For Taylor Swift, the Future of Music Is a Love Story', *Wall Street Journal*, 7 July 2014, http://www.wsj.com/articles/for-taylor-swift-the-future-of-music-is-a-love-story-1404763219.
68 David Beer, 'One-Hit Wonders Dominate as Social Media Turns Up Pace of the Pop Charts', *Conversation*, 6 Jan 2015, https://theconversation.com/one-hit-wonders-dominate-as-social-media-turns-up-pace-of-the-pop-charts-35866.
69 Ibid.
70 Interview with the author.
71 Douglas Macmillan & Greg Bensinger, 'Amazon to Buy Video Site Twitch for $970 Million', *Wall Street Journal*, 26 Aug 2014, http://www.wsj.com/articles/amazon-to-buy-video-site-twitch-for-more-than-1-billion-1408988885.
72 King, 'Candy Crush Saga Celebrates One Year Anniversary and Half a Billion Downloads', 15 Nov 2013, http://company.king.com/news-and-media/press-releases/content/press-releases/candy-crush-saga-celebrates-one-year-anniversary/.
73 Clive Thompson, '*Halo 3*: How Microsoft Labs Invented a New Science of Play', *Wired*, 21 Aug 2007, http://archive.wired.com/gaming/virtual-worlds/magazine/15–09/ff_halo.
74 Tom Chatfield, *Fun Inc*, Virgin Books, London, 2011, p. 173.
75 Sean Elder, 'A Korean Couple Let a Baby Die While They Played a Video Game', *Newsweek*, 27 July 2014, http://www.newsweek.com/2014/08/15/korean-couple-let-baby-die-while-they-played-video-game-261483.html.
76 Blake Snyder, *Save the Cat!*, Michael Wise, Studio City, 2005.
77 Interview with the author.
78 Tom Whipple, 'Slaves to the Algorithm', *Economist Intelligent Life*, May/June 2013, http://moreintelligentlife.co.uk/content/features/anonymous/slaves-algorithm.
79 Interview with the author.
80 Andrew Leonard, 'How Netflix Is Turning Viewers Into Puppets', *Salon*, 1 Feb 2013, http://www.salon.com/2013/02/01/how_netflix_is_turning_viewers_into_puppets/.
81 Tom Vanderbilt, 'Echo Nest Knows Your Music, Your Voting Choice', *Wired*, 17 Feb 2014 (chart in print version only) http://www.wired.co.uk/magazine/archive/2014/02/features/echo-nest.
82 Gabrielle Canon, 'This 28-Year-Old Knows Which Artists You'll Be Listening to 6 Months From Now', *Mother Jones*, 14 July 2014, http://www.motherjones.com/media/2014/07/can-next-big-sound-predict-future-music.
83 Leonard, 'How Netflix . . .'.

5 TOMORROW'S NEWS TODAY.

1 *The Daily Show*, 20 June 2013.
2 For details see Chris Tryhorn, '*Telegraph* Executes Saddam Blog', *Guardian*, 12 Jan 2007, http://www.theguardian.com/media/organgrinder/2007/jan/12/telegraphexecutessaddamblog.
3 Tom Standage, *The Victorian Internet*, Weidenfeld & Nicolson, London, 1998, pp. 139 & 141.
4 Nick Davies, *Flat Earth News*, Chatto & Windus, London, 2008.
5 Bill Kovach & Tom Rosenstiel, *Warp Speed*, Century Foundation, New York, 1999, p. 2.
6 Robert G. Kaiser, 'Ben Bradlee, Legendary *Washington Post* Editor, Dies at 93', 21 Oct 2014, http://www.washingtonpost.com/national/ben-bradlee-legendary-washington-post-editor-dies-at-93/2014/10/21/3e4cc1fc-c59c-11df-8dce-7a7dc354d1b1_story.html.
7 Brigid Schulte, *Overwhelmed*, Bloomsbury, London, 2014, p. 6.
8 Kovach & Rosenstiel, *Blur*, Bloomsbury, New York, 2011, p. 105.
9 American Society of News Editors, '2015 Census', 28 July 2015, http://asne.org/content.asp?pl=121&sl=415&contentid=415.
10 Nic Newman, *Digital News Report 2013*, Reuters Institute for the Study of Journalism, 2013, http://www.digitalnewsreport.org/survey/2013/executive-summary-and-key-findings-2013/.
11 Various, *The Quality and Independence of British Journalism*, Cardiff School of Journalism, 2006, p. 3.
12 Howard Rosenberg & Charles S. Feldman, *No Time To Think*, Continuum, New York, 2008, p. 26.
13 Ibid., p. 207.
14 Interview with the author.
15 The ratio of PRs to journalists in the US went from 1.2:1 in 1980 to 4:1 in 2010. Roy Greenslade, 'More PRs and Fewer Journalists Threatens Democracy', *Guardian*, 4 Oct 2012, http://www.theguardian.com/media/greenslade/2012/oct/04/marketingandpr-pressandpublishing. For more see Davies, *Flat Earth News*.
16 Alan White, Tom Phillips & Craig Silverman, 'The King of Bullshit News', *BuzzFeed*, 24 April 2015, http://www.buzzfeed.com/alanwhite/central-european-news & Alan White, 'How a Fake Viral News Story Wrecked Three People's Lives', *BuzzFeed*, 11 June 2015, http://www.buzzfeed.com/alanwhite/how-a-fake-story-ruined-three-peoples-lives.
17 See https://en.wikipedia.org/wiki/Balloon_boy_hoax.
18 Jessica Roy, 'Inside ViralNova, the Most Cynical, Amazing, Horrific, and Ingenious Media Company in New York', *New York*, 13 May 2015, http://

nymag.com/daily/intelligencer/2015/05/viralnova-new-yorks-most-cynical-media-company.html.

19 See for example http://www.dailymail.co.uk/news/article-2649381/Zoo-worker-gorilla-suit-critical-condition-shot-tranquiliser-dart-training-exercise-no-one-told-vet-wasnt-real.html.

20 Ryan Holiday, *Trust Me, I'm Lying*, Portfolio/Penguin, New York, 2013, pp. 18–29.

21 Ibid., p. 241.

22 The original thread has expired but see http://www.dailymail.co.uk/news/article-3177326/I-m-actually-bit-embarrassed-Jeremy-Corbyn-blushes-status-Mumsnet-sex-symbol-Dumbledore-sea-dog-look.html and elsewhere for the details.

23 These included the *Mail*, *Independent*, *Telegraph*, *Mirror* and *PoliticsHome*. For full details Google 'sexy Jeremy Corbyn'.

24 Kovach & Rosenstiel, *Warp Speed*, p. 97.

25 Mark Roeder, *The Big Mo*, Virgin Books, London, 2011, p. 79.

26 *The Daily Show*, 20 June 2013.

27 Jennifer Levitz, Kevin Helliker & Sara Germano, 'Deadly Blasts Rock Boston', *Wall Street Journal*, 16 April 2013, http://www.wsj.com/articles/SB10001424127887323346304578424950102614148. See 'Corrections & Amplifications' at end.

28 Oliver Willis, '*NY Post* Settles Lawsuit Over Infamous Boston Bombing "Bag Men" Cover', Media Matters, 1 Oct 2014, http://mediamatters.org/blog/2014/10/01/ny-post-settles-lawsuit-over-infamous-boston-bo/200974.

29 Ibid.

30 Alexis Madrigal, '#BostonBombing: The Anatomy of a Misinformation Disaster', *Atlantic*, April 2013, http://www.theatlantic.com/technology/archive/2013/04/-bostonbombing-the-anatomy-of-a-misinformation-disaster/275155/.

31 Alexander Abad-Santos, 'Reddit's "Find Boston Bombers" Founder Says "It Was a Disaster" but "Incredible"', *The Wire*, 22 April 2013, http://www.thewire.com/national/2013/04/reddit-find-boston-bombers-founder-interview/64455/.

32 Alex Hern, 'When Crowdsourcing Goes Wrong', *New Statesman*, 19 April 2013, http://www.newstatesman.com/world-affairs/2013/04/when-crowdsourcing-goes-wrong-reddit-boston-and-missing-student-sunil-tripathi.

33 Madrigal, 'Anatomy'.

34 Harriet Alexander, 'Tweeting Terrorism: How al Shabaab Live Blogged the Nairobi Attacks', *Telegraph*, 22 Sept 2013, http://www.telegraph.co.uk/

news/worldnews/africaandindianocean/kenya/10326863/Tweeting-terrorism-How-al-Shabaab-live-blogged-the-Nairobi-attacks.html.

35 Kashmir Hill, 'Blaming the Wrong Lanza: How Media Got It Wrong in Newtown', *Forbes*, 17 Dec 2012, http://www.forbes.com/sites/kashmirhill/2012/12/17/blaming-the-wrong-lanza-how-media-got-it-wrong-in-newtown/.

36 Roy Greenslade, '*Newsnight*'s McAlpine Scandal – 13 Days that Brought Down the BBC's Chief', *Guardian*, 19 Feb 2014, http://www.theguardian.com/media/greenslade/2014/feb/19/newsnight-lord-mcalpine.

37 Ariel Levy, 'Trial By Twitter', *New Yorker*, 5 Aug 2013, http://www.newyorker.com/magazine/2013/08/05/trial-by-twitter.

38 Graham Linehan, 'Bin Laden and *The IT Crowd*: Anatomy of a Twitter Hoax', *BBC News*, May 2011, http://www.bbc.co.uk/news/magazine-13467407. The rumour later mutated to claim that bin Laden had in fact been a fan of *The Big Bang Theory*.

39 Kovach & Rosenstiel, *Blur*, p. 39.

40 *The Onion*, 'Media Landscape Redefined by 24-Second News Cycle', 1 June 2007, http://www.theonion.com/article/media-landscape-redefined-by-24-second-news-cycle-2213.

41 Rosenberg & Feldman, *No Time*, p. 21.

42 Deborah Potter & Tom Grimes, 'Graphic Overload Hinders Understanding', *NewsLab*, Jan 2009, http://www.newslab.org/research/graphic-overload.htm.

43 Pew Research Center, *State of the Media 2013*, http://www.stateofthemedia.org/2013/special-reports-landing-page/the-changing-tv-news-landscape/#local-television-news-shrinking-pains.

44 Ibid.

45 Ibid.

46 Kovach & Rosenstiel, *Blur*, p. 130.

47 Maia Szalavitz, 'Q&A: Why "Expert" Predictions in the Media Are So Often Wrong', *Time*, 5 Aug 2011, http://healthland.time.com/2011/08/05/mind-reading-why-expert-predictions-in-the-media-are-so-often-wrong/.

48 Interview with the author.

49 Interview with the author.

50 Cates Holderness, 'What Colours Are This Dress?', BuzzFeed, 26 Feb 2015, http://www.buzzfeed.com/catesish/help-am-i-going-insane-its-definitely-blue.

51 Ravi Somaiya, 'How Facebook Is Changing the Way Its Users Consume Journalism', *New York Times*, 27 Oct 2014, http://www.nytimes.com/2014/10/27/business/media/how-facebook-is-changing-the-way-its-users-consume-journalism.html.

52 Alexandra Petri, 'Twitter, Facebook, and Ferguson – Our Awareness Problem', *Washington Post*, 18 Aug 2014, https://www.washingtonpost.com/blogs/compost/wp/2014/08/18/twitter-facebook-and-ferguson-our-awareness-problem/.
53 Nilay Patel, 'The Mobile Web Sucks', *Verge*, 20 July 2015, http://www.theverge.com/2015/7/20/9002721/the-mobile-web-sucks.
54 See the excellent 'Content Wars' series by John Herrman of the *Awl*, http://www.theawl.com/slug/the-content-wars.
55 Myles Tanzer, 'Exclusive: *New York Times* Internal Report Painted Dire Digital Picture', *BuzzFeed*, 15 May 2014, http://www.buzzfeed.com/mylestanzer/exclusive-times-internal-report-painted-dire-digital-picture.
56 Rosenberg & Feldman, *No Time*, p. 138.
57 Tom Standage, 'Bulletins from the Future', *Economist*, 7 July 2011, http://www.economist.com/node/18904136.
58 See http://www.gutenberg.org/files/45027/45027-h/45027-h.htm.
59 Emily Bell, 'Fact-Mongering Online', in Julia Hobsbawm, ed., *Where The Truth Lies*, Atlantic Books, London, 2010, p. 14.
60 Patrick Radden Keefe, 'Rocket Man', *New Yorker*, 25 Nov 2013, http://www.newyorker.com/magazine/2013/11/25/rocket-man-2.
61 Pew Research Center, *State of the Media 2013*, http://www.stateofthemedia.org/2013/special-reports-landing-page/citing-reduced-quality-many-americans-abandon-news-outlets/.
62 Interview with the author.
63 Interview with the author.
64 Interview with the author.
65 Holiday, *Trust Me*, p. 60.
66 Patrick Smith, 'This Is What It's Like to Fall in Love With a Woman who Doesn't Exist', *BuzzFeed*, 24 May 2015, http://www.buzzfeed.com/patricksmith/the-mystery-of-leah-palmer.
67 Megan Garber, 'Sit Back, Relax, and Read That Long Story – on Your Phone', *Atlantic*, 21 Jan 2014, http://www.theatlantic.com/technology/archive/2014/01/sit-back-relax-and-read-that-long-story-on-your-phone/283205/.
68 Steve Rayson, 'BuzzFeed's Most Shared Content Format Is Not What You Think', BuzzSumo, 2 May 2015, http://buzzsumo.com/blog/buzzfeeds-most-shared-content-format-is-not-what-you-think/.
69 Clive Thompson, *Smarter Than You Think*, William Collins, London, 2013, p. 235.
70 Ibid.
71 Nic Newman *et al.*, *Reuters Institute Digital News Report 2015*, Reuters Institute for the Study of Journalism, http://www.digitalnewsreport.org.

72 Eric Schmidt & Jared Cohen, *The New Digital Age*, Alfred A. Knopf, 2013, p. 48.
73 Kovach & Rosenstiel, *Blur*, p. 170.
74 See the increasing popularity of curation services such as Redef, NextDraft or even Apple's own news app.
75 See www.narrativescience.com.
76 Erin Madigan White, 'Automated Earnings Stories Multiply', *AP Blog*, 29 Jan 2015, http://blog.ap.org/2015/01/29/automated-earnings-stories-multiply/.
77 Will Oremus, 'The First News Report on the L.A. Earthquake Was Written by a Robot', *Slate*, 17 March 2014, http://www.slate.com/blogs/future_tense/2014/03/17/.
78 Kara Swisher, 'Yahoo Acquires Hipster Mobile News Reader Summly for Close to $30 Million', *AllThingsD*, 25 March 2013, http://allthingsd.com/20130325/yahoo-acquires-hipster-mobile-news-reader-summly-like-we-said-it-might/.
79 Oremus, 'The World's Smartest News Reader', *Slate*, 12 July 2013, http://www.slate.com/articles/technology/technology/2013/07/the_world_s_smartest_news_reader_prismatic_has_solved_one_of_the_internet.html.
80 Oremus, 'News Reader'.
81 See Eli Pariser, *The Filter Bubble*, Penguin, London, 2012.
82 Steven Johnson, *Future Perfect*, Allen Lane, London, 2012, p. 102.

6 THE PACE OF POLITICS

1 Foreword to Philip Gould, *The Unfinished Revolution*, Hachette, London, 2011, p. 3.
2 As found by the Obama campaign. Sasha Issenberg, *The Victory Lab*, Crown, New York, 2012, p. 285.
3 Joe Trippi, *The Revolution Will Not Be Televised*, Regan, New York, 2004, p. 37.
4 Fred Metcalf, *The Biteback Dictionary of Humorous Political Quotations*, Biteback, 2012, p. 83.
5 Michael Cockerell, 'The News from Number Ten', in Julia Hobsbawm, ed., *Where the Truth Lies*, Atlantic Books, London, 2010, p. 45.
6 Gould, *Unfinished* (first edition, 1998), p. 294.
7 Ibid.
8 Trippi, *Revolution*, p. 40.
9 Interview with the author.
10 Daniel Terdiman, 'Obama's Win a Big Vindication for Nate Silver, King of the Quants', *CNET*, 7 Nov 2012, http://www.cnet.com/uk/news/obamas-win-a-big-vindication-for-nate-silver-king-of-the-quants/.

11 Trippi, *Revolution*, p. 163.
12 Jonathan Alter, *The Promise*, Simon & Schuster, London, 2011, p. 344.
13 David Jackson, 'Obama: 24/7 Media Makes it Hard to Focus "on the Long Term"', *USA Today*, 13 Oct 2010, http://content.usatoday.com/communities/theoval/post/2010/10/obama-247-media-makes-it-hard-to-focus-on-the-long-term/1.
14 Alter, *Promise*, p. 46.
15 Tony Blair, *A Journey*, Hutchinson, London, 2010, p. 128.
16 Robert Booth, Lizzy Davies & Rajeev Syal, 'George Osborne Raises Standard in First-Class Train Row', *Guardian*, 19 Oct 2012, http://www.theguardian.com/politics/2012/oct/19/george-osborne-standard-ticket-first-class-train.
17 Interview with the author.
18 Alastair Campbell, 'Campbell on Campbell', *New Statesman*, 3 Feb 2011, http://www.newstatesman.com/uk-politics/2011/01/blair-tony-diaries-government.
19 Howard Rosenberg & Charles S. Feldman, *No Time To Think*, Continuum, New York, 2008, p. 181.
20 Ibid.
21 Interview with the author.
22 Interview with the author.
23 Interview with the author.
24 Interview with the author.
25 *Yes, Prime Minister* season 2, episode 5, 'Power to the People'.
26 Lecture to the Centre for Policy Studies, 25 Sept 2013, http://www.cps.org.uk/about/news/q/date/2013/09/25/full-text-nadhim-zahawi-mp-lecture/.
27 Interview with the author.
28 Cole Stryker, *Hacking the Future*, Overlook Duckworth, New York, 2012, p. 42.
29 Huikyong Pang, 'The 2008 Candlelight Protest in South Korea: Articulating the Paradox of Resistance in Neoliberal Globalization', Jan 2013, http://scholarcommons.usf.edu/cgi/viewcontent.cgi?article=5939&context=etd.
30 Mark Halperin & John Heilemann, *Double Down*, Penguin, New York, 2013, p. 41.
31 Peter Hennessy, *Distilling the Frenzy*, Biteback, London, 2012, p. 108.
32 Alter, *Promise*, p. 435.
33 Dana Millbank, 'America's Attention Deficit Disorder Politics', *Washington Post*, 11 July 2014, https://www.washingtonpost.com/opinions/dana-milbank-americas-attention-deficit-disorder-politics/2014/07/11/ae12bd4e-0903-11e4-bbf1-cc51275e7f8f_story.html.

34 Rosenberg & Feldman, *No Time*, p. 31.
35 Blair, *Journey*, p. 18.
36 Ibid., p. 312.
37 Interview with the author.
38 Tom Standage, *The Victorian Internet*, Weidenfeld & Nicolson, London, 1998, p. 146.
39 Ibid., p. 148.
40 Michael Lewis, 'Obama's Way', *Vanity Fair*, Oct 2012, http://www.vanityfair.com/news/2012/10/michael-lewis-profile-barack-obama.
41 Evgeny Morozov, *The Net Delusion*, Allen Lane, London, 2011, p. 267.
42 Eric Schmidt & Jared Cohen, *The New Digital Age*, Alfred A. Knopf, 2013, p. 131.
43 Thomas Friedman, lecture to Oxford Martin School, 27 April 2015, http://www.oxfordmartin.ox.ac.uk/event/2122.
44 Rosenberg & Feldman, *No Time*, pp. 187–8.
45 Yifu Dong, 'China's Clickbait Nationalism', *Foreign Policy*, 25 July 2015, https://foreignpolicy.com/2015/07/25/china-clickbait-nationalism-japan-war/.
46 Interview with the author.
47 Interview with the author.
48 Interview with the author.
49 Blair, *Journey*, p. 107.
50 Interview with the author.
51 Interview with the author.
52 Alistair Darling, *Back from the Brink*, Atlantic Books, 2011, p. 225.
53 Ibid., p. 194.
54 Interview with the author.
55 David Remnick, 'Going the Distance', *New Yorker*, 27 Jan 2014, http://www.newyorker.com/magazine/2014/01/27/going-the-distance-david-remnick.
56 Interview with the author.
57 Interview with the author.
58 Malcolm Gladwell, *Blink*, Penguin, London, 2005.
59 Ibid., pp. 72–75.
60 Charles Ballew & Alexander Todorov, 'Predicting Political Elections from Rapid and Unreflective Face Judgments', *Proceedings of the National Academy of Sciences*, Oct 2007.
61 Michael Efran & E. W. J. Patterson, 'Voters Vote Beautiful: The Effect of Physical Appearance on a National Election', *Canadian Journal of Behavioural Science*, 1974.
62 Rosenberg & Feldman, *No Time*, pp. 127–8.

63 Darling, *Brink*, p. 286.

64 Thomas Coughlan, 'How John Key Could Win the Election for David Cameron', *Cambridge Globalist*, 19 March 2015, http://cambridge-globalist.org/2015/03/19/how-john-key-could-win-election-david-cameron/.

65 Tim Dickinson, 'Roger Ailes' Keys to Campaign Success', *Rolling Stone*, 6 June 2011, http://www.rollingstone.com/politics/news/roger-ailes-keys-to-campaign-success-pictures-mistakes-and-attacks-20110606.

66 Steven Barnett, 'The Age of Contempt', *Guardian*, 28 Oct 2002, http://www.theguardian.com/politics/2002/oct/28/pressandpublishing.media.

67 Interview with the author.

68 Interview with the author.

69 'David Cameron Pledges to "Crush" Racism in Football', *BBC Sport*, 22 Feb 2012, http://www.bbc.co.uk/sport/0/football/17124938.

70 'PM Supports Weatherfield One', *BBC News*, 31 March 1998, http://news.bbc.co.uk/1/hi/uk/71934.stm.

71 Alter, *Promise*, p. 278.

72 Ibid., p. 279.

73 Ibid., p. iv.

74 Peter Hamby, 'Did Twitter Kill the Boys on the Bus?', Joan Shorenstein Center on the Press, Sept 2013, Politics and Public Policy, http://shorensteincenter.org/wp-content/uploads/2013/08/d80_hamby.pdf.

75 Ibid.

76 Dave Eggers, *The Circle*, Hamish Hamilton, London, 2013, pp. 238–9. The term is taken from Scientology, presumably to stress the Circle's cult-like qualities.

77 Jill Lepore, 'Long Division', *New Yorker*, 2 Dec 2013, http://www.newyorker.com/magazine/2013/12/02/long-division.

78 Moses Naim, 'Don't Feel Too Bad, Americans, Gridlock Is Global', *Atlantic*, 23 Oct 2013, http://www.theatlantic.com/international/archive/2013/10/don-t-feel-too-bad-americans-gridlock-is-global/280790/.

79 Interview with the author.

80 Blair, *Journey*, p. 290.

81 Ben Hammersley, *64 Things You Need to Know Now for Then*, Hodder & Stoughton, London, 2012, p. 12.

82 Alastair Campbell, 'A Technophobe No More', *Guardian*, 20 Feb 2006, http://www.theguardian.com/politics/2006/feb/20/comment.egovernment.

83 Interview with the author.

84 See Al Gore, *The Future*, WH Allen, London, 2013.

85 Interview with the author.

86 Morozov, *Delusion*, p. 239.
87 Matt Taibbi, *The Divide*, Scribe, London, 2014.
88 Daniel Goleman, *Focus*, Bloomsbury, London, 2013, p. 251.
89 John Micklethwait & Adrian Wooldridge, *The Fourth Revolution*, Allen Lane, London, 2014, p. 256.
90 Interview with the author.
91 Gore, *Future*, p. xxviii.
92 Ibid., p. 93.
93 George Packer, 'Change the World', *New Yorker*, 27 May 2013, http://www.newyorker.com/magazine/2013/05/27/change-the-world.
94 Ibid.
95 See http://www.google.co.uk/patents/US6368227.
96 'Has the Ideas Machine Broken Down?', *Economist*, 10 Jan 2013, http://www.economist.com/news/briefing/21569381-idea-innovation-and-new-technology-have-stopped-driving-growth-getting-increasing.
97 Schmidt & Cohen, *Digital Age*, p. 204.
98 Ibid., p. 205.
99 Interview with the author – Eric Schmidt.
100 Robert Colvile, 'Why the Government Has Unfriended Facebook', *Telegraph*, 27 Nov 2014, http://www.telegraph.co.uk/technology/facebook/11256524/Why-the-Government-has-unfriended-Facebook.html.
101 Brad Stone & Vernon Silver, 'Google's $6 Billion Miscalculation on the EU', *Bloomberg Businessweek*, 6 Aug 2015, http://www.bloomberg.com/news/features/2015–08–06/google-s-6-billion-miscalculation-on-the-eu.
102 Shelly Banjo, 'Uber Wins its Fight Against New York City Mayor Bill De Blasio – At Least for Now', *Quartz*, 22 July 2015, http://qz.com/461663/uber-wins-its-fight-against-new-york-city-mayor-bill-de-blasio-at-least-for-now/.
103 Oxford Martin Commission for Future Generations, 'Now for the Long Term', 18 Oct 2013, University of Oxford, http://www.oxfordmartin.ox.ac.uk/downloads/commission/Oxford_Martin_Now_for_the_Long_Term.pdf.
104 Philip Zimbardo & John Boyd, *The Time Paradox*, Rider, London, 2010, p. 181.
105 *World Economic Outlook*, IMF, April 2007, p. xv, https://www.imf.org/external/pubs/ft/weo/2007/01/pdf/exesum.pdf.
106 Interview with the author.
107 Interview with the author.
108 Trippi, *Revolution*, p. 146.
109 Douglas Carswell, *The End of Politics*, Biteback, London, 2012, p. 227.

110 See https://twitter.com/HeadUKCivServ/status/619408173980643328.
111 Carswell, *Politics*, p. 87.
112 Blair, *Journey*, pp. 18–19.
113 Andrew Haldenby, Tara Majumdar & Greg Rosen, 'Whitehall Reform: The View From the Inside', Reform, Feb 2013, http://www.reform.uk/wp-content/uploads/2014/10/Whitehall_reform_The_view_from_the_inside.pdf.
114 Ibid.
115 See for example Clayton M. Christensen, *The Innovator's Prescription*, McGraw-Hill, New York, 2009.
116 Chris Yiu with Sarah Fink, 'Smaller, Better, Faster, Stronger', Policy Exchange, 2013, p. 5 http://www.policyexchange.org.uk/images/publications/smaller%20better%20faster%20stronger.pdf.
117 Ibid., p. 36.
118 This statistic and others from the author's visit to the Government Digital Service.
119 Interview with the author.
120 Interview with the author.
121 Interview with the author.
122 Interview with the author.
123 Andrew Orlowski, 'Inside Gov.uk: "Chaos" and "Nightmare" as Trendy Cabinet Office Wrecked Govt Websites', *Register*, 18 Feb 2015, http://www.theregister.co.uk/2015/02/18/the_inside_story_of_govuk/.
124 Jon Gertner, 'Inside Obama's Stealth Startup', *Fast Company*, 15 June 2015, http://www.fastcompany.com/3046756/obama-and-his-geeks.
125 Interview with the author.
126 Yiu with Fink, 'Smaller', p. 33.
127 Ibid.
128 Ibid., p. 44.
129 Interview with the author.
130 See http://webarchive.nationalarchives.gov.uk/20130405170223/http:/www.hm-treasury.gov.uk/spend_spendingchallenge_faq.htm for details.
131 Interview with the author.
132 See https://petitions.whitehouse.gov/response/isnt-petition-response-youre-looking.
133 Patrick Kingsley, 'Participatory Democracy in Porto Alegre', *Guardian*, 10 Sept 2012, http://www.theguardian.com/world/2012/sep/10/participatory-democracy-in-porto-alegre.
134 Carswell, *Politics*, pp. 198–202.
135 Eggers, *Circle*, p. 399.
136 Carswell, *Politics*, p. 205.

137 Steven Johnson, *Future Perfect*, Allen Lane, London, 2012, p. 61.
138 Interview with the author.
139 Interview with the author.
140 Schmidt and Cohen, *Digital Age*, p. 9.
141 Micklethwait & Wooldridge, *Revolution*, p. 109.
142 Nitasha Tiku, 'Peter Thiel's Dream of a Lawless Utopia Floats On', *ValleyWag*, 23 Sept 2013, http://valleywag.gawker.com/peter-thiel-s-dream-of-a-lawless-utopia-floats-on-1368141049.
143 'Balaji Srinivasan on Silicon Valley's Ultimate Exit', SeaSteading.org, 22 Oct 2013, http://www.seasteading.org/2013/10/balaji-srinivasan-on-silicon-valleys-ultimate-exit/.
144 Oxford Martin Commission, 'Long Term'.
145 Interview with the author.
146 Ian Goldin & Mike Mariathasan, *The Butterfly Defect*, Princeton University Press, Princeton, 2014.
147 Schmidt and Cohen, *Digital Age*, p. 89.
148 Ibid., p. 157.
149 Morozov, *Delusion*, p. 272.
150 Ibid., p. 154.
151 See https://en.wikipedia.org/wiki/Office_of_Personnel_Management_data_breach.
152 John Seabrook, 'Network Insecurity', *New Yorker*, 20 May 2013, http://www.newyorker.com/magazine/2013/05/20/network-insecurity.
153 Jaron Lanier, *Who Owns the Future?*, Allen Lane, London, 2013, p. 195.

7 TIME IS MONEY.

1 Testimony to Congressional Committee on Banking, Housing, and Urban Affairs, 16 July 2002, http://www.federalreserve.gov/boarddocs/hh/2002/july/testimony.htm.
2 John Lanchester, *Whoops!: Why Everyone Owes Everyone and No One Can Pay*, Penguin, London, 2010, p. 1.
3 John Kay, 'Tailgaters Blight Markets and Motorways', *Financial Times*, 19 Jan 2010, http://www.ft.com/cms/s/0/eb1062b6–0532–11df-a85e-00144-feabdc0.html.
4 Nate Silver, *The Signal and the Noise*, Allen Lane, London, 2012, p. 368.
5 Alan Greenspan, *The Map and the Territory*, Allen Lane, London, 2013, p. 301.
6 Al Gore, *The Future*, WH Allen, London, 2013, p. 36.
7 Alex Edmans, Vivian Fang and Katharina Lewellen, 'Equity Vesting and Managerial Myopia', European Corporate Governance Institute, 12 May 2015, http://papers.ssrn.com/sol3/papers.cfm?abstract_id=2270027.

8 Carl Honoré, *The Slow Fix*, Collins, London, 2013, p. 107.
9 Kamal Ahmed, 'Davos 2011: Unilever's Paul Polman Believes We Need to Think Long Term', *Telegraph*, 15 Jan 2011, http://www.telegraph.co.uk/finance/financetopics/davos/8261178/Davos-2011-Unilevers-Paul-Polman-believes-we-need-to-think-long-term.html.
10 John Cassidy, 'The Attention-Deficit-Disorder Economy', *New Yorker*, 21 Feb 2015, http://www.newyorker.com/news/john-cassidy/twitter-buzzfeed-hurting-economic-growth.
11 Speech by Andrew Haldane in Beijing, July 2011, http://www.bis.org/review/r110720a.pdf.
12 The Kay Review of UK Equity Markets and Long-Term Thinking, Final Report, June 2012, https://www.gov.uk/government/uploads/system/uploads/attachment_data/file/253454/bis-12–917-kay-review-of-equity-markets-final-report.pdf.
13 Michael Lewis, *The Big Short*, Allen Lane, London, 2010.
14 Mark Atherton, 'The Race to be Neil Woodford's Successor', *Times*, 19 Oct 2013, http://www.thetimes.co.uk/tto/money/investment/article3898485.ece.
15 Kay Review, Final Report.
16 Ibid.
17 Philip Zimbardo & John Boyd, *The Time Paradox*, Rider, London, 2010, p. 268.
18 James Surowiecki, 'Requiem for a Dreamliner?', *New Yorker*, 4 Feb 2013, http://www.newyorker.com/magazine/2013/02/04/requiem-for-a-dreamliner.
19 Ha-Joon Chang, *23 Things They Don't Tell You About Capitalism*, Penguin, London, 2011, p. 259.
20 Ibid.
21 Lynn Parramore, 'Lazonick: How Superstar Companies Like Apple Are Killing America's High-Tech Future', Institute for New Economic Thinking, 9 Dec 2014, http://ineteconomics.org/ideas-papers/blog/how-superstar-companies-like-apple-are-killing-americas-high-tech-future.
22 See for example http://bigthink.com/videos/government-regulation-stifles-innovation.
23 Greenspan, *Territory*, p. 210.
24 Tyler Cowen, *The Great Stagnation*, Dutton, New York, 2012.
25 Rory Sutherland, 'You Can Buy Happiness', *Spectator*, 12 April 2014, http://www.spectator.co.uk/life/the-wiki-man/9180881/you-can-buy-happiness-heres-how/.
26 Clayton M. Christensen, speech to the Royal Society of Arts, 9 Sept 2013, https://www.thersa.org/events/2013/09/the-capitalists-dilemma/.

27 Susan Christopherson, 'Short-Term Profit Seeking Risks the Future of Manufacturing', 24 Sept 2013, *The Conversation*, https://theconversation.com/short-term-profit-seeking-risks-the-future-of-manufacturing-18573.
28 Ibid.
29 Ciaran Driver & Paul Temple, 'Capital Investment', UK Government Office for Science, Oct 2013, https://www.gov.uk/government/uploads/system/uploads/attachment_data/file/283884/ep8-capital-investment-trends-uk-manufacturing.pdf.
30 Andrew Ross Sorkin, *Too Big to Fail*, Penguin, London, 2010, p. 494.
31 Ibid., p. 480.
32 Iain Martin, *Making It Happen*, Simon & Schuster, London, 2013, p. 24.
33 Interview with the author.
34 Gillian Tett, *Fool's Gold*, Abacus, London, 2010, p. 74.
35 Lanchester, *Whoops!*, p. 29.
36 Michael Lewis, 'How the Eggheads Cracked', *New York Times*, 24 Jan 1999, http://www.nytimes.com/1999/01/24/magazine/how-the-eggheads-cracked.html.
37 Ibid.
38 Roger Lowenstein, *When Genius Failed*, Fourth Estate, London, 2001.
39 Interview with the author.
40 Tett, *Fool's Gold*, p. 157.
41 Interview with the author.
42 Michael Lewis, 'What Wall Street's CEOs Don't Know Can Kill You', *Bloomberg*, 26 March 2008, http://www.bloomberg.com/apps/news?pid=newsarchive&sid=aSE8yLAyALNQ.
43 Interview with the author.
44 Mark Roeder, *The Big Mo*, Virgin Books, London, 2011, p. 34.
45 See http://www.hmrc.gov.uk/manuals/cfmmanual/cfm12110.htm.
46 Interview reprinted in Lewis, ed., *Panic!*, Penguin, London, 2008, p. 117.
47 Ibid.
48 Alistair Darling, *Back from the Brink*, Atlantic Books, 2011, p. 126.
49 Jaron Lanier, *Who Owns the Future?*, Allen Lane, London, 2013, p. 68.
50 Michael Goodkin, *The Wrong Answer Faster*, Wiley, New Jersey, 2012, p. 232.
51 Ibid., p. 107.
52 Christopher Steiner, *Automate This*, Portfolio/Penguin, New York, 2012, pp. 18–48.
53 Ben Hammersley, *64 Things You Need to Know Now for Then*, Hodder & Stoughton, London, 2012, p. 149.
54 Interview with the author.

55 Robert Harris, *The Fear Index*, Hutchinson, London, 2011.
56 Goodkin, *Answer*, p. 287.
57 Steiner, *Automate*, pp. 113–20.
58 Michael Lewis, *Flash Boys*, Allen Lane, London, 2014, p. 271.
59 Andrew Haldane, speech in Beijing, July 2011.
60 Oral evidence to the Parliamentary Commission on Banking Standards, Sub-Committee G, Nov 2012, http://www.publications.parliament.uk/pa/jt201314/jtselect/jtpcbs/27/27viiI_121126g.htm (Justin Welby, the then Bishop of Durham, is also a former energy executive, so was co-opted on to the committee).
61 Interview with the author.
62 Prospectus not available online but see Harry Wilson, 'High Speed Traders Make Millions in Split Seconds', *Times*, 3 Aug 2015, http://www.thetimes.co.uk/tto/business/industries/banking/article4515384.ece.
63 Lewis, *Flash Boys.*
64 Ibid., p. 62.
65 Ibid., p. 210.
66 Ibid., p. 172.
67 Ibid., p. 266.
68 Interview with the author.
69 Interview with the author.
70 Lenterman, in interview with the author.
71 Lewis, *Flash Boys*, p. 264.
72 Patrick Hosking, 'Barclays Faces Headlong Plunge into its Dark Pool', *Times*, 27 June 2014, http://www.thetimes.co.uk/tto/business/industries/banking/article4131300.ece.
73 Harris, 'Frankenstein Finance: How Supercomputers Preying on Human Fear Are Taking Over the World's Stock Markets', *Daily Mail*, 2 Oct 2011, http://www.dailymail.co.uk/debate/article-2043943/How-supercomputers-preying-human-fear-taking-worlds-stock-markets.html.
74 Lewis, *Flash Boys*, p. 163.
75 Ibid., p. 199.
76 Ibid., p. 210.
77 Steiner, *Automate*, p. 45.
78 Jenny Strasburg & Jacob Bunge, 'Loss Swamps Trading Firm', *Wall Street Journal*, 2 Aug 2012, http://www.wsj.com/articles/SB10000872396390443866404577564772083961412.
79 Steiner, *Automate*, pp. 51–2.
80 Carol Clark, 'Controlling Risk in a Lightning-Speed Trading Environment', *Chicago Fed Letter*, 2 March 2010, https://www.chicagofed.org/publications/chicago-fed-letter/2010/march-272.

81 See for example Scott McMurray & Robert L. Rose, 'The Crash of '87', reprinted in Lewis, ed., *Panic*, p. 20.

82 See https://en.wikipedia.org/wiki/2010_Flash_Crash.

83 Philip Stafford, 'Flash Crash Explanation Questioned', *Financial Times*, 3 April 2013, http://www.ft.com/cms/s/0/52e2e7e0–9c5e-11e2–9a4b-00144feabdc0.html#axzz3hUapMkY.

84 Ben Wright, 'Can Navinder Singh Sarao Really Be a Wolf of Wall Street?', *Telegraph*, 24 April 2015, http://www.telegraph.co.uk/finance/financial-crime/11559912/Can-Navinder-Singh-Sarao-really-be-a-Wolf-of-Wall-Street.html.

85 Dave Cliff & Linda Northrup, 'The Global Financial Markets: An Ultra-Large-Scale Systems Perspective', Government Office for Science, Oct 2011, p. 4, https://www.gov.uk/government/uploads/system/uploads/attachment_data/file/289012/11–1223-dr4-global-financial-markets-systems-perspective.pdf.

86 Interview with the author.

87 Cliff & Northrup, 'Financial Markets', p. 9.

88 Dave Lauer, 'The Fragility of the US Markets', *Huffington Post*, 8 July 2012, http://www.huffingtonpost.com/dave-lauer/knight-capital-glitch_b_1738174.html.

89 Seth Stevenson, 'The Wolf of Wall Tweet', *Slate*, 20 April 2015, http://www.slate.com/articles/business/moneybox/2015/04/bot_makes_2_4_million_reading_twitter_meet_the_guy_it_cost_a_fortune.html.

90 Peter Foster, 'Bogus AP Tweet About Explosion at the White House Wipes Billions Off US Markets', *Telegraph*, 23 April 2013, http://www.telegraph.co.uk/finance/markets/10013768/Bogus-AP-tweet-about-explosion-at-the-White-House-wipes-billions-off-US-markets.html.

91 Stevenson, 'Wall Tweet'.

92 Interview with the author.

93 Gore, *Future*, p. 16.

94 Lanier, *Who Owns the Future?*, p. 69.

95 Steiner, *Automate*, p. 2.

96 Cliff & Northrup, 'Financial Markets', p. 22.

97 Clayton M. Christensen, Speech to the Royal Society of Arts, Sept 2013.

98 Ibid.

99 Kay Review, Final Report.

8 PLANET EXPRESS.

1 Sun Tzu, *The Art of Warfare*, trans. Roger T. Ames, Ballantine, New York, 1993, p. 120.

2 See for example http://www.spacex.com/hyperloop.
3 Tom Vanderbilt, *Traffic*, Allen Lane, London, 2008, p. 133.
4 Texas Transportation Institute, *2015 Urban Mobility Scorecard*, http://d2dtl5nnlpfror.cloudfront.net/tti.tamu.edu/documents/mobility-scorecard-2015.pdf.
5 European Commission figures, via http://ec.europa.eu/transport/themes/urban/urban_mobility/index_En.htm.
6 CEBR, '50% Rise in Gridlock Costs By 2030', 14 Oct 2014, http://www.cebr.com/reports/the-future-economic-and-environmental-costs-of-gridlock/.
7 Christopher Hope, 'Get Used to Driving at 40mph, Says Top Highways Official', *Telegraph*, 13 May 2014, http://www.telegraph.co.uk/motoring/news/10825851/Motorway-speeds-Get-used-to-driving-at-40mph-says-top-highways-official.html.
8 Vanderbilt, *Traffic*, p. 17.
9 Daniel Sperling & Deborah Gordon, 'Two Billion Cars', *Transportation Research News*, Dec 2008, http://onlinepubs.trb.org/onlinepubs/trnews/trnews259billioncars.pdf.
10 See for example Vanderbilt, *Traffic*, p. 113.
11 See https://en.wikipedia.org/wiki/China_National_Highway_110_traffic_jam.
12 Andreas Schäfer, 'Long-Term Trends in Global Passenger Mobility', *Papers from the 12th US Frontiers of Engineering*, Winter 2006.
13 Office for National Statistics, 'Commuting and Personal Well-Being, 2014', http://www.ons.gov.uk/ons/dcp171766_351954.pdf.
14 Ibid.
15 Frank Partnoy, *Wait*, PublicAffairs, Philadelphia, 2012, p. 202.
16 Texas A&M Transportation Institute, '2012 Urban Mobility Report', http://mobility.tamu.edu/ums/report/.
17 Vanderbilt, *Traffic*, p. 132.
18 Interview with the author.
19 Vanderbilt, *Traffic*, p. 138.
20 Ibid., p. 114.
21 'Smart Traffic Signals', Carnegie Mellon University, http://www.cmu.edu/homepage/computing/2012/fall/smart-traffic-signals.shtml.
22 Jeff Stibel, *Breakpoint*, Palgrave Macmillan, New York, 2013, p. 126.
23 Jeremy Rifkin, *The Zero Marginal Cost Society*, Palgrave Macmillan, New York, 2014, p. 226.
24 Ibid., p. 227.
25 Burkhard Bilger, 'Auto Correct', *New Yorker*, 25 Nov 2013, http://www.newyorker.com/magazine/2013/11/25/auto-correct.

26 Ibid.
27 Ibid.
28 Nick Reed, UK Transport Research Laboratory, in interview with the author.
29 Bilger, 'Auto Correct'.
30 Ibid.
31 Interview with the author.
32 Ibid.
33 Leonard E. Read, 'I, Pencil', http://www.econlib.org/library/Essays/rdPncl1.html.
34 Thomas L. Friedman, *The World is Flat*, Farrar, Straus and Giroux, New York, 2005.
35 Toby Poston, 'Thinking Inside the Box', *BBC News*, 25 April 2006, http://news.bbc.co.uk/1/hi/business/4943382.stm.
36 Gaia Vince, *Adventures in the Anthropocene*, Chatto & Windus, London, 2014, p. 320.
37 Steve Hilton with Jason & Scott Bade, *More Human*, WH Allen, London, 2015, p. 170.
38 James Surowiecki, 'Uber Alles', *New Yorker*, 16 Sept 2013, http://www.newyorker.com/magazine/2013/09/16/uber-alles-2.
39 James Gleick, *Faster*, Abacus, London, 2000, pp. 75–6.
40 Farhad Manjoo, 'I Want It Today', *Slate*, 11 July 2012, http://www.slate.com/articles/business/small_business/2012/07/amazon_same_day_delivery_how_the_E_commerce_giant_will_destroy_local_retail_.html.
41 Andrew Griffin, 'Amazon Introduces One Hour Delivery with Amazon Prime Now', *Independent*, 18 Dec 2014, http://www.independent.co.uk/life-style/gadgets-and-tech/news/amazon-introduces-one-hour-delivery-with-amazon-prime-now-9933146.html.
42 Alexis Madrigal, 'Inside Google's Secret Drone-Delivery Program', *Atlantic*, 28 Aug 2014, http://www.theatlantic.com/technology/archive/2014/08/inside-googles-secret-drone-delivery-program/379306/.
43 Ben Popper, 'Amazon Provides New Details on its Plan for a Drone Superhighway in the Sky', *Verge*, 28 July 2015, http://www.theverge.com/2015/7/28/9058211/amazon-new-details-plan-delivery-drone.
44 Michael Hobbes, 'The Myth of the Ethical Shopper', *Huffington Post*, 16 July 2015, http://highline.huffingtonpost.com/articles/en/the-myth-of-the-ethical-shopper/.
45 Brad Stone, *The Everything Store*, Transworld, London, 2014, p. 237.
46 Ken Silverstein, *The Secret World of Oil*, Verso, London, 2014, p. 102.
47 Ibid., p. 103.
48 Paul McMahon, *Feeding Frenzy*, Profile, London, 2013, p. 166.

49 Ibid., p. 145.
50 Ibid., p. 152.
51 Ibid., p. 48.
52 Evan Osnos, *Age of Ambition*, Bodley Head, London, 2014, p. 4.
53 'In China, An Unprecedented Economic Transformation', Yale School of Management, 11 May 2015, http://som.yale.edu/news/2015/05/china-unprecedented-economic-transformation.
54 Osnos, *Ambition*, p. 60.
55 Ibid., p. 24.
56 Ibid., p. 41.
57 Ibid., p. 58.
58 Ibid., p. 64.
59 John Micklethwait & Adrian Wooldridge, *The Fourth Revolution*, Allen Lane, London, 2014, p. 4.
60 Gwynn Guilford, 'Everything You've Heard About China's Stock Market Crash Is Wrong', *Quartz*, 27 Aug 2015, http://qz.com/486476/everything-youve-heard-about-chinas-stock-market-crash-is-wrong/.
61 Patrick French, *India*, Allen Lane, London, 2011, p. 197.
62 Ibid., p. 204.
63 Oliver Balch, *India Rising*, Faber, London, 2012, p. 92.
64 United Nations, 'A World of Cities', Aug 2014, http://www.un.org/en/development/desa/population/publications/pdf/popfacts/PopFacts_2014–2.pdf.
65 Ibid.
66 Balch, *India Rising*, p. 3.
67 Ian Johnson, 'As Beijing Becomes a Supercity, the Rapid Growth Brings Pains', *New York Times*, 19 July 2015, http://www.nytimes.com/2015/07/20/world/asia/in-china-a-supercity-rises-around-beijing.html.
68 McKinsey Global Institute, 'Preparing for China's Urban Billion', Feb 2009, http://www.mckinsey.com/insights/urbanization/preparing_for_urban_billion_in_china.
69 P. D. Smith, *City*, Bloomsbury, London, 2012, p. 312.
70 The global population is believed to have passed 7 billion in March 2012. For the latest estimate see http://www.worldometers.info/world-population/.
71 Mike Davis, *Planet of Slums*, Verso, London, 2007, p. 192.
72 John Arlidge, 'The Instant City', *Sunday Times*, 7 Sept 2014, http://www.thesundaytimes.co.uk/sto/Magazine/article1452143.ece.
73 PwC, 'Tailwinds 2015', June 2015, http://www.pwc.com/en_US/us/industrial-products/publications/assets/pwc-tailwinds-rising-passenger-demand.pdf.

74 'Energy Use in Cities', TheWorldIsUrban.com, citing World Bank statistics, March 2011, http://theworldisurban.com/2011/03/energy-use-in-cities/.
75 Arlidge, 'Instant City'.
76 Jonah Lehrer, 'A Physicist Solves the City', *New York Times*, 19 Dec 2010, http://www.nytimes.com/2010/12/19/magazine/19Urban_West-t.html.
77 Ibid.
78 Ibid.
79 Asian Development Bank, 'Asia 2050: Realizing the Asian Century', Aug 2011, http://www.iopsweb.org/researchandworkingpapers/48263622.pdf.
80 PwC, 'The World in 2050', Feb 2015, http://www.pwc.com/gx/en/issues/the-economy/assets/world-in-2050-february-2015.pdf.

9 RACING TO DESTRUCTION.

1 Paul Roberts, *The End of Food*, Bloomsbury, London, 2008, p. 71.
2 Diane Ackerman, *The Human Age*, Headline, London, 2014, p. 17.
3 Mark Lynas, *The God Species*, Fourth Estate, London, 2011, p. 6.
4 Ackerman, *Human Age*, p. 42.
5 Roberts, *Food*, p. 89.
6 Lecture attended by the author.
7 Roberts, *Food*, p. 71.
8 Charles Duhigg, *The Power of Habit*, William Heinemann, London, 2012, p. 27.
9 Roberts, *Food*, p. 43.
10 Ibid.
11 Maggie Jackson, *Distracted*, Prometheus Books, New York, 2009, p. 103.
12 Roberts, *Food*, p. 97.
13 Ibid., p. 43.
14 Carolyn Steel, *Hungry City*, Chatto & Windus, London, 2008, p. 166.
15 Ibid.
16 Roberts, *Food*, p. 50.
17 Steve Hilton with Jason & Scott Bade, *More Human*, WH Allen, London, 2015, p. 124.
18 Roberts, *Food*, p. 30.
19 Ibid., p. 41.
20 Ibid., p. 51.
21 Graham Kendall, 'The Science that Makes Us Spend More in Supermarkets, and Feel Good While We Do It', *The Conversation*, 4 March 2014, https://theconversation.com/the-science-that-makes-us-spend-more-in-supermarkets-and-feel-good-while-we-do-it-23857.

22 Deena Shankar, '19 Supermarket Mind Games that Get You to Buy More Junk Food', *BuzzFeed*, 18 June 2014, http://www.buzzfeed.com/deenashanker/ways-supermarkets-trick-you-into-buying-more-junk.
23 Steel, *Hungry*, p. 269.
24 Roberts, *Food*, p. 65.
25 Dana Gunders, 'Wasted', Natural Resources Defence Council, Aug 2012, http://www.nrdc.org/food/files/wasted-food-ip.pdf.
26 Steel, *Hungry*, p. 159.
27 Ibid., p. 63.
28 Roberts, *Food*, p. 65.
29 Steel, *Hungry*, p. 60.
30 Ibid.
31 Ibid., p. 59.
32 Ibid., p. 101.
33 Lecture attended by the author.
34 Steel, *Hungry*, p. 101.
35 Ibid., p. 102.
36 Andrew Brown, 'In a Nutshell: How Glorious Bread Is Made', *Telegraph*, 12 June 2013, http://blogs.telegraph.co.uk/news/andrewmcfbrown/100221472/in-a-nutshell-how-glorious-bread-is-made/.
37 Roberts, *Food*, p. 46.
38 Ibid., p. 49.
39 Paul McMahon, *Feeding Frenzy*, Profile, London, 2013, pp. 21–2.
40 Ibid., p. 22.
41 Roberts, *Food*, pp. 3–5.
42 Ibid., p. 69.
43 See for example Tim Spector, *The Diet Myth*, Weidenfeld & Nicolson, London, 2015, or Alanna Collen, *10% Human*, William Collins, London, 2015.
44 Philip Lymbery with Isabel Oakeshott, *Farmageddon*, Bloomsbury, London, 2014, p. 192.
45 Roberts, *Food*, p. 76.
46 Ibid.
47 Lymbery with Oakeshott, *Farmageddon*, p. 35.
48 Ibid., p. 84.
49 Ibid., p. 93.
50 Steve Hilton *et al.*, *More Human*, p. 170.
51 Roberts, *Food*, p. 63.
52 Steel, *Hungry*, p. 96.
53 Ibid.
54 Roberts, *Food*, p. 194.

55 Ibid., p. 72.
56 Lymbery with Oakeshott, *Farmageddon*, p. 165.
57 Ibid.
58 Ibid.
59 Roberts, *Food*, p. 176.
60 James R. Johnson *et al.*, 'Antimicrobial-Resistant and Extraintestinal Pathogenic *Escherichia coli* in Retail Foods', *Journal of Infectious Diseases*, April 2005, http://www.ncbi.nlm.nih.gov/pubmed/15747237.
61 McMahon, *Frenzy*, p. 136.
62 Roberts, *Food*, p. 77.
63 Lily Kuo, 'The World Eats Cheap Bacon at the Expense of North Carolina's Rural Poor', *Quartz*, 14 July 2015, http://qz.com/433750/the-world-eats-cheap-bacon-at-the-expense-of-north-carolinas-rural-poor/.
64 Lymbery with Oakeshott, *Farmageddon*, p. 13.
65 Roberts, *Food*, p. 229.
66 'World Without Water', *Spiegel Online*, 13 Aug 2015, http://www.spiegel.de/international/world/global-water-shortage-exacerbated-by-droughts-and-misuse-a-1047527.html.
67 Roberts, *Food*, p. 229.
68 McMahon, *Frenzy*, p. 66.
69 Lymbery with Oakeshott, *Farmageddon*, p. 83.
70 Ibid., p. 83.
71 Thomas Friedman, lecture to Oxford Martin School, 27 April 2015, http://www.oxfordmartin.ox.ac.uk/event/2122.
72 Fred Pearce, *Peoplequake*, Eden Project Books, London, 2010, p. 239.
73 Evan D. G. Fraser & Andrew Rimas, *Empires of Food*, Random House, London, 2010, p. 38.
74 Lynas, *God Species*, p. 194.
75 Christopher Hooton, 'Oxford Street Is the Most Polluted Place in the World, Say Scientists', *Independent*, 7 July 2014, http://www.independent.co.uk/news/uk/oxford-street-is-the-most-polluted-place-in-the-world-say-scientists-9589276.html.
76 Patti Waldmeir, 'Shanghai Notebook: Bronchial Set Seek Blue Sky Breathing in China', *Financial Times*, 24 Nov 2014, http://www.ft.com/cms/s/0/9a68bf4c-731f-11e4–907b-00144feabdc0.html.
77 Lynas, *God Species*, p. 207.
78 Ibid., p. 63.
79 Brian Clark Howard, 'Worst Drought in 1,000 Years Predicted for American West', *National Geographic*, 12 Feb 2015, http://news.nationalgeographic.com/news/2015/02/150212-megadrought-southwest-water-climate-environment/.
80 Fraser & Rimas, *Empires*, p. 8.

81 Pearce, *Peoplequake*, p. 241 (average figure from http://www.sustainablescale.org/conceptualframework/understandingscale/measuringscale/ecologicalfootprint.aspx).

82 The difference comes from whether you account for carbon emissions. See Charlotte McDonald, 'How Many Earths Do We Need?', 16 June 2015, *BBC News*, http://www.bbc.co.uk/news/magazine-33133712.

83 Gaia Vince, *Adventures in the Anthropocene*, Chatto & Windus, London, 2014, p. 103.

84 As Fred Pearce says, rising consumption rather than population has been responsible for almost all of our additional ecological impact over the past 30 years. Pearce, *Peoplequake*, p. 241.

85 George Monbiot, *Bring on the Apocalypse*, Atlantic/Guardian, London, 2009, p. 6.

86 Roberts, *The End of Oil*, Bloomsbury, London, 2005, p. 223.

87 Dale Allen Pfeiffer, cited in McMahon, *Frenzy*, p. 85.

88 Raj Patel, *The Value of Nothing*, Portobello, London, 2009, p. 164.

89 Roberts, *Oil*, p. 241.

90 'Special Report: Feeding the World', *Economist*, 24 Feb 2011, http://www.economist.com/node/18200702.

91 McMahon, *Frenzy*, p. 94.

92 Ibid., p. 78.

93 Ibid.

94 Ibid., p. 79.

95 Ibid., p. 81.

96 Pearce, *Peoplequake*, p. 257.

97 McMahon, *Frenzy*, p. 90.

98 Roberts, *Food*, p. 286.

99 Ibid., p. 285.

100 Ibid., p. 286.

101 Steel, *Hungry*, p. 315.

102 Jeff Stibel, *Breakpoint*, Palgrave Macmillan, New York, 2013, p. 33.

103 Lymbery with Oakeshott, *Farmageddon*, p. 267.

104 Ibid., p. 285.

105 Ibid.

106 Ackerman, *Human*, p. 88.

107 Ibid., p. 123.

108 Tracy McVeigh, 'Humble Spud Poised to Launch a World Food Revolution', *Observer*, 18 Oct 2014, http://www.theguardian.com/science/2014/oct/18/humble-potato-poised-to-launch-food-revolution.

109 Alex Kirby, 'Heat-Beater Beans Could Help Keep Millions Fed', *Climate News Network*, 26 March 2015, http://www.climatenewsnetwork.net/heat-tolerant-beans-could-help-keep-millions-fed/.

110 Roberts, *Food*, p. 240.
111 Vince, *Adventures*, p. 290.
112 Deborah Netburn, 'Scientists Develop a Tomato Plant that Can Grow All Night', *Los Angeles Times*, 5 Aug 2014, http://www.latimes.com/science/sciencenow/la-sci-sn-continuous-light-tolerant-tomato-plants-20140805-story.html.
113 Alok Jha, 'Google's Sergey Brin Bankrolled World's First Synthetic Beef Hamburger', *Guardian*, 5 Aug 2013, http://www.theguardian.com/science/2013/aug/05/google-sergey-brin-synthetic-beef-hamburger.
114 Jeffrey Towson, 'Why Jack Ma Becoming a Dairy Farmer is a Big Deal', *South China Morning Post*, 2 Sept 2014, http://www.scmp.com/comment/blogs/article/1583478/why-jack-ma-becoming-dairy-farmer-big-deal.
115 Ibid.
116 Andrew Griffin, 'Why Is Toshiba Growing No-Soil Veg in its Old Factories?', *Independent*, 17 Nov 2014, http://www.independent.co.uk/life-style/gadgets-and-tech/why-is-toshiba-growing-nosoil-veg-in-its-old-factories-9865180.html.
117 Kevin Gray, 'MIT's Incubator-Grown Plants Might Hold Key to Food Crisis', *Wired*, 6 Oct 2014, http://www.wired.co.uk/magazine/archive/2014/10/features/server-farm.
118 Ibid.
119 Olivia Solon, 'Vast Underground Bomb Shelter Reappropriated by Urban Farmers', *Wired*, 11 Feb 2014, http://www.wired.co.uk/news/archive/2014–02/11/underground-farm-zero-carbon-food.
120 Lizzie Widdicombe, 'The End of Food', 12 May 2014, *The New Yorker*, http://www.newyorker.com/magazine/2014/05/12/the-end-of-food.
121 Tim Bradshaw, 'Food 2.0', *Financial Times*, 31 Oct 2014, http://www.ft.com/cms/s/2/bfa6fca0–5fbb-11e4–8c27–00144feabdc0.html.
122 Daniel Fromson, 'The Silicon Valley Race to Build a Fake-Meat Burger That Just Might Save the World', *Grub Street*, 2 June 2015, http://www.grubstreet.com/2015/06/silicon-valley-fake-meat-burger.html.
123 Bradshaw, 'Food 2.0'.
124 Lecture attended by the author.
125 Peter Marsh, *The New Industrial Revolution*, Yale University Press, London, 2012, p. 142.
126 Lynas, *God Species*, p. 240.
127 Vince, *Adventures*, p. 320.
128 Lecture by Tim Kruger, James Martin Fellow, Oxford Geoengineering Programme, Oxford Martin School.

129 For a good summary of the basic approaches and their probable effects see Royal Society, 'Geoengineering the Climate: Science, Governance and Uncertainty', Sept 2012, https://royalsociety.org/policy/publications/2009/geoengineering-climate/.
130 Vince, *Adventures*, p. 295.
131 Adam Piore, 'Rise of the Insect Drones', *Popular Science*, 29 Jan 2014, http://www.popsci.com/article/technology/rise-insect-drones.
132 See OECD, 'Effective Carbon Prices', 4 Nov 2013, http://www.oecd-ilibrary.org/environment/effective-carbon-prices_9789264196964-en.
133 Lynas, *God Species*, p. 241.

CONCLUSION – FAST FORWARD.

1 Kurt Vonnegut, *Player Piano*, Dial Press, New York, 2006 ed., pp. 289–90.
2 David Leonhardt, 'Amtrak Crash and America's Declining Construction Spending', *New York Times*, 13 May 2015, http://www.nytimes.com/2015/05/14/upshot/amtrak-crash-and-americas-declining-construction-spending.html?rref=upshot&abt=0002&abg=1.
3 Taken from the *CIA World Factbook*, https://www.cia.gov/library/publications/the-world-factbook/fields/2177.html.
4 Fred Pearce, *Peoplequake*, Eden Project Books, London, 2010, p. 296.
5 Robert M. Sapolsky, *Monkeyluv*, Vintage, London, 2006, pp. 199–208.
6 Adam Rutherford, *Creation: The Future of Life*, Viking, London, 2013, p. 11.
7 See J. Craig Venter, *Life at the Speed of Light*, Abacus, London, 2013.
8 Rutherford, *Creation*, p. 56.
9 Andy Greenberg, 'This Is The World's First Entirely 3D-Printed Gun', *Forbes*, 3 May 2013, http://www.forbes.com/sites/andygreenberg/2013/05/03/this-is-the-worlds-first-entirely-3d-printed-gun-photos/.
10 Andrew Hessel, Mark Goodman & Stephen Kotler, 'Hacking the President's DNA', *Atlantic*, Nov 2012, http://www.theatlantic.com/magazine/archive/2012/11/hacking-the-presidents-dna/309147/.
11 Amy Maxmen, 'The Genesis Engine', *Wired*, July 2015, http://www.wired.com/2015/07/crispr-dna-editing-2.
12 Nick Bostrom, *Superintelligence*, Oxford University Press, Oxford, 2014, p. 44.
13 Sian Westlake, ed., *Our Work Here Is Done*, Nesta, London, 2014, p. 26.
14 Hans Greimel, 'Honda's New Plant Takes Manufacturing to the Next Level', *Auto News*, 1 Dec 2013, http://www.autonews.com/article/20131201/OEM01/312029986/hondas-new-plant-takes-manufacturing-to-the-next-level.

15 Michael Kan, 'Foxconn Expects Robots to Take Over More Factory Work', *PC World*, 27 Feb 2015, http://www.pcworld.com/article/2890032/foxconn-expects-robots-to-take-over-more-factory-work.html.
16 Andy Larson, 'Rethink's Baxter vs Universal Robots: Which Collaborative Robot Is Best for You?', *Machine & Factory Automation*, http://cross-automation.com/blog/rethinks-baxter-vs-universal-robots-which-collaborative-robot-best-you.
17 Kurzweil, *How to Create a Mind*, Duckworth Overlook, London, 2012, p. 117.
18 Lecture attended by the author.
19 Kurzweil, *Mind*, p. 279.
20 Ibid., p. 281.
21 See http://cser.org.
22 Steve Connor, 'American Scientists Controversially Recreate Deadly Spanish Flu Virus', *Independent*, 11 June 2014, http://www.independent.co.uk/news/science/american-scientists-controversially-recreate-deadly-spanish-flu-virus-9529707.html.
23 Bostrom, *Superintelligence*, p. 123.
24 Ibid., p. 97.
25 Ibid., p. 70.
26 Ibid., p. 65.
27 Ibid., p. 259.
28 Ibid., p. vii.
29 Demonstration witnessed by the author.
30 See for example Sarah Buhr, 'CVS Health Taps IBM's Watson to Predict Patient Health Decline Before It Happens', *TechCrunch*, 30 July 2015, http://techcrunch.com/2015/07/30/cvs-health-taps-ibms-watson-to-help-predict-patients-health-decline-before-it-happens/
31 Kevin Kelly, 'The Three Breakthroughs That Have Finally Unleashed AI on the World', *Wired*, 27 Oct 2014, http://www.wired.com/2014/10/future-of-artificial-intelligence/

Index

A Note on the Author

Robert Colvile is a political commentator and technology writer whose work appears in the *Telegraph*, *Financial Times*, *Spectator*, Politico, *CapX* and elsewhere. He was previously news director at BuzzFeed UK and comment editor at the *Daily Telegraph*, where he also worked as a columnist and leader writer as well as overseeing the paper's science pages. He has a Masters degree from Cambridge in International Relations, has been a regular pundit on Sky News, and is a global fellow at PS21, the Project for the Study of the 21st Century. He is also the author of an influential report on how the internet is transforming British politics, which was praised by Chancellor George Osborne among others.

robertcolvile.com / @rcolvile

A Note on the Type

The text of this book is set in Bembo, which was first used in 1495 by the Venetian printer Aldus Manutius for Cardinal Bembo's *De Aetna*. The original types were cut for Manutius by Francesco Griffo. Bembo was one of the types used by Claude Garamond (1480–1561) as a model for his Romain de l'Université, and so it was a forerunner of what became the standard European type for the following two centuries. Its modern form follows the original types and was designed for Monotype in 1929.